hematics

SpringerBriefs in Mathematics showcases expositions in all areas of mathematics and applied mathematics. Manuscripts presenting new results or a single new result in a classical field, new field, or an emerging topic, applications, or bridges between new results and already published works, are encouraged. The series is intended for mathematicians and applied mathematicians.

For further volumes:
http://www.springer.com/series/10030

SpringerBriefs in Mat

Björn Birnir

The Kolmogorov-Obukhov Theory of Turbulence

A Mathematical Theory of Turbulence

Springer

Björn Birnir
University of California
Department of Mathematics
Santa Barbara, CA 93106
USA

ISSN 2191-8198 ISSN 2191-8201 (electronic)
ISBN 978-1-4614-6261-3 ISBN 978-1-4614-6262-0 (ebook)
DOI 10.1007/978-1-4614-6262-0
Springer New York Heidelberg Dordrecht London

Library of Congress Control Number: 2012956056

Mathematics Subject Classification (2010): 60H15, 35R60, 76B79, 35Q35, 35Q30, 76D05, 37L55

Printed on acid-free paper

Springer is part of Springer Science+Business Media (www.springer.com)

To Inga, Adda and Einar.

Preface

In this book we present the recently developed statistical theory of turbulence in a form that can be appreciated by physicists, mathematicians, and engineers. The theory is grounded in probability theory and we develop from probability all the results that are necessary to understand the turbulence theory, without proofs. However, references are given to standard texts where the proofs and more background details can be found. The goals are to find estimates for structure functions of turbulence that are realized in simulation and experiments; to derive the invariant measure of turbulence both for the one-point statistics and the two-point statistics; and finally to derive the probability density function (PDF) for the statistics that are used in practice. We do not assume any mathematical background but familiarity with basic probability theory and partial differential equations obviously helps.

We will see that the Navier–Stokes equation for all but the largest scales in turbulent flow can be expressed as a stochastic Navier–Stokes equation (1.65). The stochastic forcing results from instabilities of the flow that magnifies small ambient noise and saturates its growth into large stochastic forcing. This has been modeled before by a Reynolds decomposition and by a coarse graining of the flow. The stochastic force is generic and is determined by the general principles of probability with a minimum of physical inputs. It consists of two components: additive noise and multiplicative noise and the additive component is determined by the central limit theorem and the large deviation principle. The physical input is that these two terms must produce similar scalings because they are the detailed description of the same dissipative processes. This determines the rate in the large deviation principle. The multiplicative noise multiplies the fluid velocity and models jumps (vorticity concentrations) in the velocity gradient. It is expressed by a generic Poisson process where only the rate needs to be given. This rate is determined by the spectral analysis of the (linearized) Navier–Stokes operator and the requirement, following [64], that the dimension of the most singular vorticity structure (filaments) is one. Thus the stochastic forcing is generic and determined with two mild physical inputs.

The stochastic Navier–Stokes equation can be expressed as an integral equation (2.17) and the log-Poissonian processes found by She and Leveque and explored by She and Waymire and Dubrulle are produced from the multiplicative noise by the Feynman–Kac formula. This gives a satisfying mathematical derivation of the intermittency phenomena that had earlier been derived from impirical considerations. Moreover, the integral equations show how the Navier–Stokes evolution and the log-Poissonian intermittency processes act on the dissipation processes to produce the intermittency in the dissipation. This is a mathematical derivation of the experimental observation that intermittent dissipation processes accompany intermittent velocity variations. Using the integral equation, we get an estimate on all the structure functions of the velocity differences in turbulence. The evidence from simulations and experiments is that this upper bound is reached in turbulent flow. Why the inertial cascade achieves this maximal efficiency in the energy transfer remains to be explained.

We then build on Hopf's [29] ideas to compute the invariant measure of turbulent flow. This measure can be computed because it solves a linear functional differential equation, the Kolmogorov–Hopf equation; see [56]. It turns out to be an infinite-dimensional Gaussian multiplied by a (discrete) Poisson distribution. This Poisson distribution corresponds to the intermittency and the log-Poisson processes. Then by taking the trace of the invariant measure we get the PDF of the velocity differences. We first derive the functional differential equation (PDE) for the PDF and then show that there are infinitely many PDFs, each corresponding to a particular moment because of the intermittency corrections. The PDE (3.15) for the sequence of PDFs can also be solved and the PDFs turn out to be the normalized inverse Gaussian (NIG) distributions of Barndorff–Nielsen [7]. Their parameters are easly computed and we see how to do this for both simulations and experiments.

It is interesting to notice that although the solution of the Navier–Stokes equation may not be unique or smooth the invariant measure of the velocity differences (3.12) is still well defined by Leray's [42] existence theory. Moreover, different velocities produce equivalent measures, so the statistical observables of turbulence are unique although the turbulent velocity may not be.

The theory presented in this book must be complemented by a dynamical systems theory for the large-scale structures in fluid flow and eventually one wants to work out how the small-scale flow presented here influences the large-scale dynamics. This is a material for future research, but hopefully the tools presented in this book will also be helpful in that endeavor.

Santa Barbara, California, United States — Björn Birnir

Acknowledgements The author would like to acknowledge a large number of colleagues with whom the results in this book have been discussed and who have provided valuable insights. They include Ole Barndorff-Nilsen and Jurgen Schmiegel in Aarhus, Henry McKean, K. R. Sreenivasan, and R. Varadhan in New York, Z.-S. Zhe in Beijing, Ed Waymire in Oregon, E. Bodenschatz and H. Xu in Gottingen, Michael Wilczek in Munster and M. Sørensen in Copenhagen. He also benefitted from conversations with J. Peinke, M. Oberlack, E. Meiburg, B. Eckhardt, S. Childress, L. Biferale, L-S. Yang, A. Lanotto, K. Demosthenes, M. Nelkin, A. Gylfason, V. L'vov, and many others. This research was supported in part by the The Courant Institute and the Project of Knowledge Innovation Program (PKIP) of Chinese Academy of Sciences, Grant No. KJCX2.YW.W10, whose support is gratefully acknowledged and by a grant from the UC Santa Barbara Academic Senate.

Contents

Chapter 1
The Mathematical Formulation of Fully Developed Turbulence

1.1 Introduction to Turbulence

The purpose of the research discussed in this book is to develop new mathematical tools that open the theory of turbulence up to theoretical investigations. Great strides are currently being made both in turbulence experiments and simulations, but the new mathematical development will allow theoreticians to compare with both simulations and experiments and make new predictions useful to both areas.

The ultimate goal of turbulence research is to develop methods to systematically improve the simulations of turbulent systems. Such methods have been ad hoc so far, with different techniques applied to each situations. The new theory will permit a systematic approach where the simulations and experiment can be made increasingly accurate in a stepwise fashion.

These developments will eventually have a big effect on technology permitting improvements in aircraft and car design, more efficient travel in and out of space, less pollution, more fuel efficiency, and greater efficiency of wind turbines and wave energy farms. It will help to understand weather patterns and greatly advance weather predictions. In addition it will aid a wide variety of applications of turbulence in industry and science.

In 1941 Kolmogorov and Obukhov [34, 35, 49] proposed a statistical theory of turbulence based on dimensional arguments. The main consequence and test of this theory was that the structure functions of the velocity differences of a turbulent fluid

$$E(|u(x,t)-u(x+l,t)|^p) = S_p = C_p l^{p/3}$$

should scale with the distance (lag variable) l between them, to the power $p/3$. This theory was immediately criticized by Landau for not taking into account the influence of the large flow structure on the constants C_p and later for not including the influence of the intermittency, in the velocity fluctuations, on the scaling exponents.

In 1962 Kolmogorov and Obukhov [36, 50] proposed a corrected theory where both of those issues were addressed. They also pointed out that the scaling exponents

B. Birnir, *The Kolmogorov-Obukhov Theory of Turbulence: A Mathematical Theory of Turbulence*, SpringerBriefs in Mathematics, DOI 10.1007/978-1-4614-6262-0_1,

for the first two structure functions could be corrected by log-normal processes. For higher-order structure functions the log-normal processes gave intermittency corrections inconsistent with contemporary simulations and experiments.

In [18] the author showed how the central limit theorem and large deviation principle produce additive noise that must be added to the Navier–Stokes equation for a proper description of fully developed turbulence. In addition, jumps in the velocity gradients produce multiplicative noise that must also be added to the deterministic equation. Following Hopf's work in 1952 [29], he used this stochastic Navier–Stokes equation to compute the scaling of the structure functions and compute the probability density function (PDF) of the velocity differences. The Feynman–Kac formula produces log-Poisson processes from the stochastic Navier–Stokes equation. These processes, first found by She and Leveque [64], Waymire [65], and Dubrulle in 1995 [23], give the correct intermittency corrections to the structure functions of turbulence.

The PDF of the velocity differences (two-point statistics) turned out to be the generalized hyperbolic distribution first suggested by Barndorff-Nielsen in 1977 [6]. The author compared the theoretical PDF with PDFs obtained from DNS simulations and wind tunnel experiments and found excellent agreement [18].

In this book the Kolmogorov–Obukhov statistical theory of turbulence, with intermittency corrections, is derived from a stochastic Navier–Stokes equation with generic noise. Various aspects of the theory are proven and its extension to turbulent vorticity developed. In collaboration with both experimental groups and researchers doing direct Navier–Stokes simulations (DNS) the theory is verified and used to predict both the experimental and numerical results.

The literature on turbulence is vast and we will not attempt to survey it here. However, we will mention some important texts and recent papers that have addressed different aspects of the theory. First the two encyclopedic books by Momin and Yaglom [46, 47] laid the foundation for much of the later research and set the stage for the questions that needed to be answered. In particular they stressed Kolmogorov's point of view that the theory of turbulence was statistical and the ultimate goal was to find the invariant measure of turbulence. Then the invariant measure should be used to show that, if large-scale structures that characterize different flows are removed, the turbulent flow is uniquely ergodic in a statistical sense.

The earlier mathematical books on the subject, [5, 61, 74], emphasized the theory of finite-dimensional attractors, see also [25], and tried to build the turbulence theory on them. There seems to be fundamental reasons why this cannot work. The attractors of dissipative nonlinear partial differential equations (PDEs) on compact domains are finite-dimensional, see [62, 63], and using ideas of Milnor, see [31, 45], it can show that the core of the attractors that attracts a (prevalent) set of full measure in phase space is typically low-dimensional. Thus these attractors cannot be used to describe fully developed turbulence that is for all practical purposes infinite-dimensional and continues to explore the full phase space for all time. In fact, the role of the attractor is played by the invariant measure for turbulence systems and this invariant measure is supported on the whole infinite-dimensional space. The techniques developed by the above authors have, however, proven to be very useful in the development of the stochastic theory of turbulence presented in this book.

The physics and engineering literature has been more faithful, with the exception of [74], to the infinite-dimensionality and ergodicity of turbulent flow, see [10, 26, 55], and the 1-point statistics were essentially worked out by Batchelor [9] and Townsend [72] using the symmetries of homogeneous and isotropic turbulence. Good reviews can be found in [48] and [68]. The theory by Kraichnan of passive scalar is a milestone in turbulence theory, see [20], and a guiding light for later developments; see also his other works [37–39]. The multi-fractal theory detailed in [26] is able to reproduce many results from experiments and simulations. However, it involves a large number of parameters and is not fully satisfying from a mathematical perspective.

Although it has been understood from the time of Taylor [70] that the theory of turbulence is statistical it is only recently that serious attempts have been made to develop the stochastic theory of turbulence. This theory was initiated for nonlinear PDEs by Sinai [67] and much of it has been developed by him with his students and collaborators. The stochastic forcing that is added to the deterministic Navier–Stokes equation models the influence of the random environment on the fluid in fully developed turbulence. Some of the authors who have developed this theory are Da Prato and Zabczyk [57, 58], Flandoli and Gatarek [24], Kuksin and Shirikyan [41], Mattingly and Hairer [27, 28], and Debussche and Odasso [21], to name a few. The results have mostly been for the two-dimensional stochastic Navier–Stokes equation and/or noise that is white both in time and space.

In retrospect it is clear why more progress has not made on the stochastic theory of turbulence until [18]. There were three problems blocking the way: Firstly, the nature of the noise was not understood. It cannot be white but is in fact a homogeneous Lévy process, see Sect. 1.7.1; this form of the noise was inspired by McKean [44]. Secondly, the mistaken belief that the uniqueness of the three-dimensional solutions had to be known for the invariant measure to exist. In fact, the existence of the measure only requires Leray's theory; see Sect. 4.1. Thirdly, the theory of infinite-dimensional Ito processes was not developed until the book [56] by Da Prado appeared in 2006 and only then could it be adapted to the stochastic Navier–Stokes equation, as will be explained in this book.

We will use the Reynolds decomposition below to decompose the flow into large-scale flow and small-scale flow describing Kolmogorov's inertial range; see Sect. 1.7. The deterministic Navier–Stokes equation describes the laminar small-scale flow; it is unstable at large Reynolds numbers because of small noise present in any fluid; see Sect. 1.4. The existence of the invariant measure for the laminar flow is easily proven, see [25], but it can also be computed. It is a special case of the measure in Theorem 3.1, when E, Q, and $p^k_{m_k} \to 0$. We have been told after a visitor to Kolmogorov in the 1980s, that Kolmogorov had then already computed this invariant measure, or the solution of Hopf's equation, and was very disappointed to find it to be trivial. Namely, the invariant measure for the small-scale laminar flow is a delta function centered at the origin, which is a trivial stationary solution. Thus the deterministic Navier–Stokes equation (1.1) describes laminar flow and the stochastic Navier–Stokes equation (1.65) describes fully developed turbulent flow. The statistical theory of this turbulent flow is explained in this book.

1.2 The Navier–Stokes Equation for Fluid Flow

Fluid flow is described by the deterministic Navier–Stokes equation

$$u_t + u \cdot \nabla u = \nu \Delta u - \nabla p \qquad (1.1)$$
$$u(x,0) = u_0(x)$$

with the incompressibility conditions

$$\nabla \cdot u = 0, \qquad (1.2)$$

where $u(x), x \in \mathbb{R}^3$, is the velocity of the fluid and ν is the kinematic viscosity. Eliminating the pressure p using (1.2) gives the equation

$$u_t + u \cdot \nabla u = \nu \Delta u + \nabla \{\Delta^{-1}[\text{trace}(\nabla u)^2]\}. \qquad (1.3)$$

The pressure is eliminated by taking the divergence of (1.1) using the vector identity

$$\nabla \cdot (u \cdot \nabla u) = \text{trace}(\nabla u)^2 + u \cdot \nabla(\nabla \cdot u) = \text{trace}(\nabla u)^2$$

by condition (1.2). This gives the equation

$$\Delta p = \text{trace}(\nabla u)^2 \qquad (1.4)$$

for the pressure that is solved with periodic or Neumann boundary conditions. The solvability condition

$$\int_D \nabla \cdot (u \cdot \nabla u) \mathrm{d}x = \int_{\partial D} (u \cdot \nabla u) \cdot n \, \mathrm{d}\sigma = 0$$

is clearly satisfied with periodic boundary conditions on a torus $D = \mathbb{T}^3$, where $\mathrm{d}\sigma$ denotes the surface element. In the case of Neumann boundary conditions $f = \frac{\partial p}{\partial n}$, on ∂D, must also satisfy the solvability condition

$$\int_{\partial D} f \mathrm{d}\sigma = \int_{\partial D} (u \cdot \nabla u) \cdot n \, \mathrm{d}\sigma.$$

If these conditions are satisfied and the forcing $\text{trace}(\nabla u)^2$ reasonably smooth, the pressure equation (1.4) has a unique solution $p = \Delta^{-1}[\text{trace}(\nabla u)^2]$.

The turbulence of the fluid is quantified by the dimensionless Reynolds number $Re = \frac{UL}{\nu}$ where U is a typical velocity of the flow and L is a typical length scale associated with the flow. A more physically relevant Reynolds number is $R_\lambda = \frac{U\lambda}{\nu}$, where λ is the Taylor correlation length in turbulence. A rule of thumb for moderate Reynolds number is $R_\lambda \sim \sqrt{15Re}$. Thus the transition to turbulence occurs at $R_\lambda \sim 100$, flows are typically fully turbulent at $R_\lambda \sim 200$, and a small stream can have Reynolds number $R_\lambda \sim 400$ and a large river $R_\lambda \sim 4{,}000$; see [16].

The deterministic Navier–Stokes equation describes laminar flow that may exist when the Reynolds number is large, but then laminar flow is usually unstable. Small noise prevalent in nature is magnified by the instabilities in the flow and it becomes more useful to consider the velocity $u(x)$ in turbulent flow to be a stochastic process; see [35]. Then u satisfies a stochastic Navier–Stokes equation

$$\begin{aligned} du &= (\nu\Delta u - u\cdot\nabla u + \nabla\{\Delta^{-1}[\text{trace}(\nabla u)^2]\})dt + df_t, \\ u(x,0) &= u_0(x). \end{aligned} \tag{1.5}$$

Here df_t denotes the stochastic forcing in fully developed turbulence.

Much effort has gone into trying to derive the form of the stochastic forcing df_t in the stochastic Navier–Stokes equation (1.5) for particular cases of fluid flow and flow boundaries. Most of these efforts have been in vain because the noise in fully developed turbulence does not seem to care how it arose, at least not sufficiently far away from the boundary. Instead the noise seems to take a general form, depending only on small environmental noise that was magnified by the fluid instabilities and this growth then saturated by the nonlinearities present in the flow (and in the Navier–Stokes equation); see [15]. Below we will assume that the stochastic forcing has a general form stipulated by probability theory and use this form and the structure of the Navier–Stokes equation to derive the invariant measure and the PDF for turbulence. Then we will compare this PDF with PDFs obtained from simulations and fluid experiments.

1.2.1 Energy and Dissipation

If we let D denote the volume in space and put periodic or vanishing (no-slip) velocity boundary condition on the boundary ∂D, then we derive a differential equation relating the mean energy and the mean enstrophy:

$$\mathscr{E} = \frac{1}{2|D|}\int_D |u(x,t)|^2 dx, \quad \mathscr{C} = \frac{1}{2|D|}\int_D |\nabla u(x,t)|^2 dx. \tag{1.6}$$

Here $|D|$ denotes the volume of D and "mean" refers to the fact that we are dividing the energy and enstrophy by the volume. Multiplying (1.1) by u and integrating over D we get, by integration by parts,

$$\frac{d}{dt}\mathscr{E} = -2\nu\mathscr{C},$$

because all the other terms integrate to zero by the vanishing boundary conditions. This equation will play a big role in the Leray theory in Sect. 4.1. Similarly, one can get the equation

$$\frac{d}{dt}\mathscr{H} = -2\nu\mathscr{H}_\omega,$$

where $\omega = \nabla \times u$ is the vorticity and

$$\mathscr{H} = \frac{1}{2|D|}\int_D u\cdot\omega(x,t)\mathrm{d}x, \quad \mathscr{H}_\omega = \frac{1}{2|D|}\int_D \omega\cdot\nabla\times\omega(x,t)\mathrm{d}x, \tag{1.7}$$

are called, respectively, the mean helicity and mean vortical helicity; see [26]. We now define the mean energy dissipation

$$\varepsilon = -\frac{\mathrm{d}}{\mathrm{d}t}\mathscr{E}. \tag{1.8}$$

It will play a central role in the statistical theory in Chap. 2.

1.3 Laminar Versus Turbulent Flow

It has been clear to investigators of fluid flow at least from the time of the great hydraulic engineer Leonardo da Vinci (1452–1519) that the flow has two forms: one regular and another highly irregular. In the words of another engineer Osborne Reynolds in 1883, see [60],

> The internal motion of water assumes one or other of two broadly distinguishable forms-either the elements of the fluid follow one another along lines of motion which lead in the most direct manner to their destination or they eddy about in sinuous paths the most indirect possible.

In modern terminology these flows are called respectively laminar and turbulent and it was O. Reynolds who realized that which kind of flow one observed was a question of the ratio of the inertial and viscous forces. He also identified the dimensionless parameter, expressing this ratio, the Reynolds number,

$$Re = \frac{UL}{\nu} \tag{1.9}$$

that determined whether the flow is laminar or turbulent. Here U is a typical velocity of the flow, L is a typical length scale in the flow, and ν is the kinematic viscosity. Reynolds also found the onset of turbulence experimentally to be at $Re = 500$ and the fluid to be fully turbulent at $Re = 2{,}000$. In fluid flow in a pipe the length scale L is the diameter of the pipe and for flow in a channel L is the depth or width of the channel. But for isotropic or homogeneous flow L is not obviously determined by the physical domain of flow and then it makes sense to use the more general Taylor–Reynolds number

$$R_\lambda = \frac{U\lambda}{\nu},$$

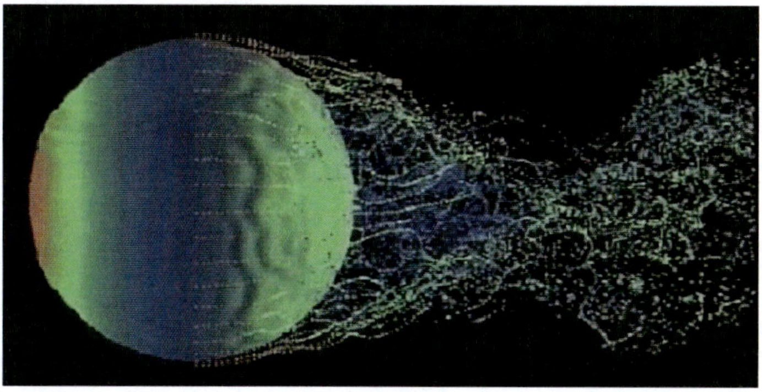

Fig. 1.1 A sphere in laminar flow with a turbulent wake.

where λ is the typical correlation length in the flow; see Sect. 1.2. In fully developed turbulent flow, the turbulence is not uniformly distributed throughout the flow but forms turbulent structures that are interspersed by laminar flow. This is called intermittency. As stated by Marusic and Nickels in [20]:

> Spatial intermittency of the small-scale motion refers to the fact that the dissipation of energy (and other characteristics of the flow) is not uniformly distributed throughout the flow, as had been previously thought, but instead occurs in very intense events that are sparsely distributed. The spatial means values are then averages of very large, very rare events.

Intermittency of turbulence was already noticed by Reynolds and investigated in great detail by, for example, Townsend [72]. He considered, in [71], models of the fine-scale motion consisting of

> a random distribution of vortex sheets and lines, in which the vorticity distribution is effectively stationary in time, due to balance between the opposing effects of vorticity diffusion by molecular viscosity, and vorticity production and convection by the turbulent shear

Using that the skewness of the velocity fluctuations is empirically negative, Townsend showed that a random array of sheets is more likely than a random array of vortex lines. For more along these lines, see [79, 80].

Kraichnan [40] pointed out that there are several different sources of intermittency. First, non-Gaussian statistics of the energy-containing scales (large eddies) can influence the statistics of the inertial range. Second, the intermittency can build up in the inertial range cascade. Thirdly, intermittency effects intrinsic to the dissipation range can also influence the inertial range statistics. At finite Reynolds numbers these different sources of intermittency may be impossible to separate.

The laminar solution may persist after the Reynolds number has become so high that the flow is fully turbulent. But then it has become unstable and is physically irrelevant. We will illustrate this with examples in the next section.

1.4 Two Examples of Fluid Instability Creating Large Noise

Consider the Navier–Stokes equation (1.1) describing turbulent flow in the center of a wide and deep river. We consider the flow to be in a box representing a typical stretch of the river and impose periodic boundary conditions on the box. Since we are mostly interested in what happens in the direction along the river we take our x-axis to be in that direction.

We will assume that the river flows fast and pick an initial condition of the form

$$U(0) = U_0 e_1, \tag{1.10}$$

where U_0 is a large constant and e_1 is a unit vector in the x direction. Clearly this initial condition is not sufficient because the fast flow may be unstable and the white noise ubiquitous in nature will grow into small velocity and pressure oscillations; see, for example, [11]. But we perform a thought experiment where white noise is introduced into the fast flow at $t = 0$. This experiment may be hard to perform in nature, but it is easily done numerically. It means that we should look for a solution of the form

$$U(x,t) = U_0 e_1 + u(x,t), \tag{1.11}$$

where $u(x,t)$ is smaller than U_0 but not necessarily small. However, in a small initial interval $[0,t_0]$, u is small and satisfies (1.5) linearized about the fast flow U_0:

$$\begin{aligned} u_t + U_0 \partial_x u &= \Delta u + \dot{f}, \\ u(x,0) &= 0, \end{aligned} \tag{1.12}$$

$\dot{f} = \frac{\mathrm{d}f}{\mathrm{d}t}$, driven by the noise

$$f = \sum_{k \neq 0} c_k^{1/2} b_t^k e_k.$$

The $e_k = \mathrm{e}^{2\pi \mathrm{i} k \cdot x}$ are (three-dimensional) Fourier components and each comes with its own independent Brownian motion b_t^k. None of the coefficients of the vectors $c_k^{1/2} = (c_1^{1/2}, c_2^{1/2}, c_3^{1/2})$ vanish because the turbulent noise was seeded by truly white noise (white both in space and in time). f is not white in space because the coefficients $c_k^{1/2}$ must have some decay in k so that the noise term in (1.12) makes sense. However, to determine the decay of the $c_k^{1/2}$s will now be a part of the problem. The form of the turbulent noise f expresses the fact that in turbulent flow there is a continuous source of small white noise that grows and saturates into turbulent noise that drives the fluid flow. The decay of the coefficients $c_k^{1/2}$ expresses the spatial coloring of this larger noise in turbulent flow. We have set the kinematic viscosity ν equal to one for computational convenience, but it can easily be restored in the formulas.

This modeling of the noise is the key idea that makes everything else work. The physical reasoning is that the white noise ubiquitous in nature grows into the noise f that is characteristic for turbulence and the differentiability properties of the turbulent velocity u are the same as those of the turbulent noise.

The justification for considering the initial value problem (1.12) is that for a short time interval $[0,t_0]$ we can ignore the nonlinear terms

$$-u\cdot\nabla u+\nabla\{\Delta^{-1}[\text{trace}(\nabla u)^2]\}$$

in (1.5). But this is only true for a short time t_0; after this time we have to start with the solution of (1.12):

$$u_0(x,t)=\sum_{k\neq 0} c_k^{1/2}\int_0^t e^{(-4\pi^2|k|^2+2\pi i U_0 k_1)(t-s)}db_s^k e_k(x) \tag{1.13}$$

as the first iterate in the integral equation

$$u(x,t)=u_0(x,t)+\int_{t_0}^t e^{K(t-s)} * [-u\cdot\nabla u+\nabla\Delta^{-1}(\text{trace}(\nabla u)^2)]ds, \tag{1.14}$$

where e^{Kt} is the (oscillatory heat) kernel in (1.13). In other words to get the turbulent solution we must take the solution of the linear equation (1.12) and use it as the first term in (1.32). It will also be the first guess in a Picard iteration. The solution of (1.12) can be written in the form

$$u_0(x,t)=\sum_{k\neq 0} c_k^{1/2}A_t^k e_k(x),$$

where the

$$A_t^k=\int_0^t e^{(-4\pi^2|k|^2+2\pi i U_0 k_1)(t-s)}db_s^k \tag{1.15}$$

are independent Ornstein–Uhlenbeck processes with mean zero; see, for example, [58].

Now it is easy to see that the solution of the integral equation (1.14) $u(x,t)$ satisfies the driven Navier–Stokes equation:

$$\begin{aligned} u_t+U_0\partial_x u &= \Delta u-u\cdot\nabla u+\nabla\Delta^{-1}(\text{trace}(\nabla u)^2)+\sum_{k\neq 0} c_k^{1/2}\dot{b}_t^k e_k, \quad t>t_0,\\ u_t+U_0\partial_x u &= \Delta u+\sum_{k\neq 0} c_k^{1/2}\dot{b}_t^k e_k, \quad u(x,0)=0, \quad t\le t_0, \end{aligned} \tag{1.16}$$

and the above argument is the justification for studying the initial value problem (1.16). We will do so numerically in the next section. The solution u of (1.16) still satisfies the periodic boundary conditions and the incompressibility condition

$$\nabla\cdot u=0. \tag{1.17}$$

The mean of the solution u_0 of the linear equation (1.12) is zero by the formula (1.13) and this implies that the solution u of (1.16) also has mean zero:

$$\bar{u}(t)=\int_{\mathbb{T}^3} u(x,t)dx=0. \tag{1.18}$$

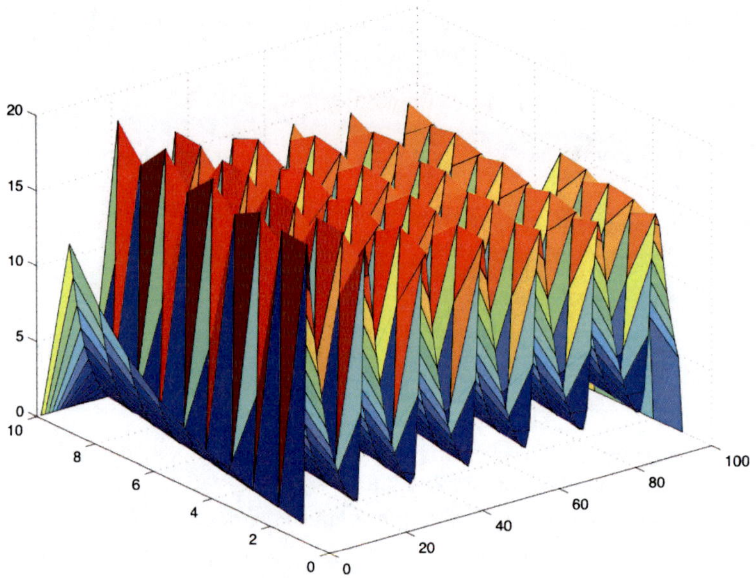

Fig. 1.2 The traveling wave solution of the heat equation for the flow velocity $U_0 = 85$. The perturbations are frozen in the flow. The x-axis is space, the y-axis time, and the z-axis velocity u.

1.4.1 Stability

The uniform flow $U = U_0 e_1$ seems to be a stable solution of (1.12) judging from the solution (1.13). Namely, all the Fourier coefficients are decaying. However, this is deceiving; first the Brownian motion b_t^k is going to make the amplitude of the kth Fourier coefficient large in due time with probability one. More importantly if U_0 is large then (1.12) has traveling wave solutions that are perturbations "frozen in the flow," and for U_0 even larger these traveling waves are unstable and start growing. For U_0 large enough this happens after a very short initial time interval and makes the flow immediately become fully turbulent. The role of the white noise is then not to cause enough growth eventually for the nonlinearities to become important, but rather to immediately pick up (large) perturbations that grow exponentially. These are the large fluctuations that are observed in most turbulent flows. In Fig. 1.1, we show the traveling wave solution of the transported heat equation (1.12), with $U_0 = 85$. In Fig. 1.2, where the flow has increased to $U_0 = 94$, the traveling wave has become unstable and grows exponentially. Notice the difference in vertical scale between the figures.

Thus the white noise grows into a traveling wave that grows exponentially. This exponential growth is saturated by the nonlinearities and subsequently the flow becomes turbulent. This is the mechanism of explosive growth of turbulence of a uniform stream; see Fig. 1.3.

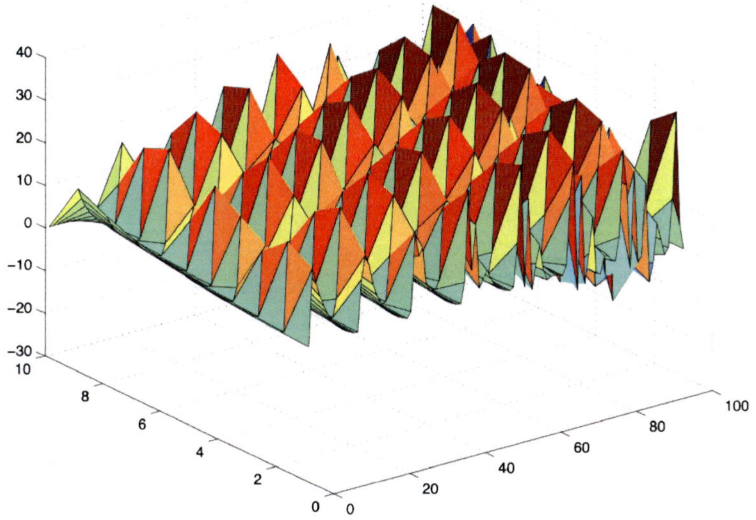

Fig. 1.3 The traveling wave solution of the heat equation for the flow velocity $U_0 = 94$. The perturbations are growing exponentially. The x-axis is space, the y-axis time, and the z-axis velocity u.

Now we turn from the previous numerical example to an analytic one. We can assume that the velocity of the flow is of the form

$$U = \mathbf{U} + u,$$

where $\mathbf{U}$ is a prescribed flow (vector) and establishes the existence of the correction u which constitutes the turbulent part of the velocity. This is a perturbative approach but u is not necessarily small. It can typically be as large, but not larger, as $\mathbf{U}$.

We will denote the mean flow in the fully developed turbulent state by U_1 and assume that uniform flow with rotation is of the form

$$\frac{\partial x}{\partial t} = \mathbf{U} = U_1 j_1 - A\sin(\Omega t + \theta_0) j_2 + A\cos(\Omega t + \theta_0) j_3, \tag{1.19}$$

where the rotation can be extended in a periodic fashion from $\mathbb{T}^3$ to $\mathbb{R}^3$.[1] One can also extend a convection cell pattern from four copies of $\mathbb{T}^3$ to $\mathbb{R}^3$ and we will use that below. This implies that the deterministic particle motion in the rotating uniform flow is simply

$$x(t) = [U_1 j_1 + \frac{A}{\Omega}\cos(\Omega t + \theta_0) j_2 + \frac{A}{\Omega}\sin(\Omega t + \theta_0) j_3]. \tag{1.20}$$

[1] For physical applications, see [30], cylindrical coordinates are more appropriate but cumbersome.

By the same reasoning as above we can choose the coordinates so that the mean flow component $U_1 j_1$ (1.19) is in the x_1 direction and this direction is the axis of the rotation.

First consider the stirred Navier–Stokes equation

$$\begin{aligned} U_t + U\cdot\nabla U &= \nu\Delta U - \nabla\Delta^{-1}\text{trace}(\nabla U)^2 \\ &\quad -A\Omega\cos(\Omega t+\theta_0)j_2 - A\Omega\sin(\Omega t+\theta_0)j_3 \\ U(x,0) &= U_1 j_1 - A\sin(\theta_0)j_2 + A\cos(\theta_0)j_3, \end{aligned} \tag{1.21}$$

where we have used the incompressibility conditions

$$\nabla\cdot U = 0 \tag{1.22}$$

to eliminate the pressure term. We want to consider turbulent flow driven by a unidirectional mean flow and to do that we consider the flow to be in a box and impose periodic boundary conditions on the box. Since we are mostly interested in what happens in the direction along the unidirectional flow we take our x_1-axis to be in that direction. The source of the small (white) noise can be thought of as fluctuations in the stirring rate of the uniform flow in (1.21).

The corresponding stochastic Navier–Stokes equation can be written as

$$\begin{aligned} du &= (\nu\Delta u - U_1\partial_{x_1}u + A\sin(\Omega t+\theta)\partial_{x_2}u - A\cos(\Omega t+\theta)\partial_{x_3}u \\ &\quad -u\cdot\nabla u + \nabla\Delta^{-1}[\text{trace}(\nabla u)^2])dt + \sum_{k\neq 0} c_k^{1/2} db_t^k e_k, \end{aligned} \tag{1.23}$$

where

$$\begin{aligned} &\frac{\partial u}{\partial t} + U_1\partial_{x_1}u - A\sin(\Omega t+\theta)\partial_{x_2}u + A\cos(\Omega t+\theta)\partial_{x_3}u + u\cdot\nabla u \\ &\quad = \nu\Delta u + \nabla\Delta^{-1}[\text{trace}(\nabla u)^2] \end{aligned} \tag{1.24}$$

is the driven Navier–Stokes equation (1.21) for $u = U - U_1 j_1 + A\sin(\Omega t+\theta)j_2 - A\cos(\Omega t+\theta)j_3$. $U_1 j_1$ is now the constant mean flow of the (fully developed) turbulent fluid and $\sum_{k\neq 0} c_k^{1/2} db_t^k e_k$ models the noise in fully developed turbulent flow. We will take the initial condition to be zero, $u(x,0)=0$, for convenience and assume that the incompressibility condition

$$\nabla\cdot u(x,t) = 0$$

is satisfied. However, the problem is just as easily solved with a nontrivial initial condition; see Theorem 4.3.

The first question one might ask about (1.23) is how the noise got introduced into the equation; see [15]. To answer that question consider the Navier–Stokes equation

$$U_t + U\cdot\nabla U = \nu\Delta U + \nabla\{\Delta^{-1}[\text{trace}(\nabla U)^2]\} \tag{1.25}$$

and linearize it about the divergence-free initial flow $\mathbf{U} = U_0 j_1 + U'(x_1, -\frac{x_2}{2}, -\frac{x_3}{2})^{\mathrm{T}}$. Here T denotes transpose and $\mathbf{U}$ is construed to be the periodic extension of the above formula from $\mathbb{T}^3$ to $\mathbb{R}^3$:

$$\begin{aligned} u_t + U_0 \partial_{x_1} u + U' \begin{pmatrix} u_1 \\ -\frac{u_2}{2} \\ -\frac{u_3}{2} \end{pmatrix} + U' \begin{pmatrix} x_1 \\ -\frac{x_2}{2} \\ -\frac{x_3}{2} \end{pmatrix} \cdot \nabla u + U' U_0 j_1 + (U')^2 \begin{pmatrix} x_1 \\ \frac{x_2}{4} \\ \frac{x_3}{4} \end{pmatrix} \\ = \nu \Delta u + \nabla \Delta^{-1} \left(\frac{3}{2} U'^2 + 2U' (\partial_{x_1} u_1 - \partial_{x_2} u_2 - \partial_{x_3} u_3) \right), \qquad (1.26) \\ u(x,0) = 0. \end{aligned}$$

We assume that there is small noise

$$\mathrm{d}f^0 = \sum_{k \neq 0} h_k^{1/2} \mathrm{d}b_t^k e_k \qquad (1.27)$$

present in the fluid. Note that the coefficients $h_k^{1/2} \neq c_k^{1/2}$ are small. Then u satisfies the linear stochastic PDE:

$$\begin{aligned} \mathrm{d}u = \Bigg[\nu \Delta u - U_0 \partial_{x_1} u - U' \begin{pmatrix} u_1 \\ -\frac{u_2}{2} \\ -\frac{u_3}{2} \end{pmatrix} - U' \begin{pmatrix} x_1 \\ -\frac{x_2}{2} \\ -\frac{x_3}{2} \end{pmatrix} \cdot \nabla u - U' U_0 j_1 \\ -(U')^2 \begin{pmatrix} x_1 \\ \frac{x_2}{4} \\ \frac{x_3}{4} \end{pmatrix} \cdot \nabla u + \nabla \Delta^{-1} \left(\frac{3}{2} U'^2 + 2U' (\partial_{x_1} u_1 - \partial_{x_2} u_2 - \partial_{x_3} u_3) \right) \Bigg] \mathrm{d}t \\ + \sum_{k \neq 0} h_k^{1/2} \mathrm{d}b_t^k e_k, \qquad (1.28) \end{aligned}$$

where the term $\sum_{k \neq 0} h_k^{1/2} \mathrm{d}b_t^k e_k$ represents stochastic forcing by the small ambient noise.

The solution of this linear equation can be found by use of a Fourier series and it is

$$\begin{aligned} u(x,t) = \sum_{k \neq 0} \int_0^t \mathrm{e}^{-(4\nu\pi^2 |k|^2 + 2\pi \mathrm{i} U_0 k_1)(t-s)} \\ \times \left(h_k^{1/2}(1) \mathrm{e}^{-U'(t-s)} j_1 + h_k^{1/2}(2) \mathrm{e}^{\frac{U'}{2}(t-s)} j_2 + h_k^{1/2}(3) \mathrm{e}^{\frac{U'}{2}(t-s)} j_3 \right) \mathrm{d}b_t^k e_k \\ + O(|U'|), \end{aligned}$$

where $h_k^{1/2}(i), i = 1,2,3$, denotes the ith entry of the three vectors $h_k^{1/2}$. Now the expectation of $u(x,t)$ vanishes, but the variation is

$$E(|u|_2^2)(t) = \sum_{k\neq 0} \int_0^t e^{-8\nu\pi^2|k|^2(t-s)} \times \left(c_k(1)e^{-2U'(t-s)} + c_k(2)e^{U'(t-s)} + c_k(3)e^{U'(t-s)}\right) ds + O(|U'|^2). \tag{1.29}$$

This shows that on one hand the small noise will grow exponentially in time, in the $e_k j_1$ direction, if

$$U' < 0 \tag{1.30}$$

and if $|U'| > 8\pi^2\nu|k|^2$ for some $k \in \mathbb{Z}^3 \setminus \{0\}$, but $|U'|$ is small compared to the exponentially growing term. If on the other hand

$$U' > 0 \tag{1.31}$$

the small noise will grow exponentially in the $e_k j_2$ and $e_k j_3$ directions (in function space), again with $|U'|$ small compared to the exponentially growing term.

The exponential growth of the noise will, however, only continue for a limited time. The growth is quickly saturated by the nonlinear terms in the equation and fluid becomes fully turbulent.

The initial value problem (1.23) can also be written as an integral equation

$$u(x,t) = u_0(x,t) - \int_0^t e^{K(t-s)} * (u\cdot\nabla u - \nabla\Delta^{-1}[\text{trace}(\nabla u)^2])ds, \tag{1.32}$$

where e^{Kt} is the (oscillatory heat) kernel

$$e^{Kt} * f = \sum_{k\neq 0} \int_0^t e^{-(4\pi^2|k|^2+2\pi i U_1 k_1)(t-s)-2\pi i A(k_2,k_3)[\sin(\Omega t+\theta)-\sin(\Omega s+\theta)]} \hat{f}(k,s) ds\, e_k, \tag{1.33}$$

$A(k_2,k_3) = A\sqrt{k_2^2+k_3^2}$, $\theta = \tan^{-1}(\frac{k_2}{k_3}) - \theta_0$ and

$$u_0(x,t) = \sum_{k\neq 0} c_k^{1/2} \int_0^t e^{-(4\pi^2|k|^2+2\pi i U_1 k_1)(t-s)-2\pi i A(k_2,k_3)[\sin(\Omega t+\theta)-\sin(\Omega s+\theta)]} db_s^k e_k(x) \tag{1.34}$$

is a sum of independent oscillatory processes,

$$A_t^k = \int_0^t e^{-(4\pi^2|k|^2+2\pi i U_1 k_1)(t-s)-2\pi i A(k_2,k_3)[\sin(\Omega t+\theta)-\sin(\Omega s+\theta)]} db_s^k \tag{1.35}$$

with mean zero; see, for example, [58]. These processes are reminiscent of Ornstein–Uhlenbeck processes and we will call them oscillatory Ornstein–Uhlenbeck-type processes below.

The mean (average) of the solution u_0 of the linear equation is zero by the formula (1.34) and this implies that the solution u of (1.23) also has mean (average) zero:

$$\bar{u}(t) = \int_{\mathbb{T}^3} u(x,t)\mathrm{d}x = 0. \tag{1.36}$$

It also implies that

$$|U|_2^2 = |\mathbf{U}|^2 + |u|_2^2 \tag{1.37}$$

for $U = \mathbf{U} + u$ and $\mathbf{U} = U_1 j_1 - A\sin(\Omega t + \theta_0)j_2 + A\cos(\Omega t + \theta_0)j_3$ with $|\mathbf{U}| = \sqrt{U_1^2 + A^2}$. We will derive a priori estimates for U in Sect. 4.1 but then apply them to u in Sects. 4.2–4.3 using (1.37).

1.5 The Central Limit Theorem and the Large Deviation Principle, in Probability Theory

We will now start our excursion into probability with the aim of introducing the reader to the basic concepts but referring to standard texts for all the details. A triple $(\Omega, \mathscr{F}, \mathbb{P})$ is called a probability space where Ω is a set, $\mathscr{F}$ is a sigma algebra of the subsets (events) of Ω, and $\mathbb{P}$ is a probability measure. An $\mathscr{F}$ measurable function $X : \Omega \to \mathbb{R}^n$ is called a random variable and any random variable induces a probability measure on $\mathbb{R}^n$:

$$\mu_X(B) = \mathbb{P}(X^{-1}(B)), \quad B \subset \mathbb{R}^n,$$

called the distribution of X. For more information about probability spaces and random variables, see [13, 14, 51].

The central limit theorem and the strong law of large numbers are a refinement of the weak law of large numbers; see [14]. The weak law of large numbers says that if

$$M_n = \frac{X_1 + X_2 + \cdots + X_n}{n}$$

is the average of independent and identically distributed random variables X_k, $k = 1, \ldots, n$, on a probability space $(X, \mathscr{F}, \mathbb{P})$, then

$$\lim_{n \to \infty} \mathbb{P}(|M_n - m| \geq \varepsilon) = 0,$$

where $m = E(X_k)$, the common mean of the random variables, is assumed to exist. The strong law of large numbers is a stronger version of this statement, namely,

Theorem 1.1. *If the X_k, $k = 1, \ldots, n$ are independent and identically distributed random variables and $m = E(X_k)$, then*

$$\mathbb{P}(\lim_{n\to\infty} M_n = m) = 1.$$

For proof see [14], where the following example is also presented.

Example 1.1. The strong law of large number for Bernoulli trials (flipping of a biased coin) is the classical example. Then $\mathbb{P}(X_k = 1) = p$, $\mathbb{P}(X_k = 0) = 1 - p$, $E(X_k) = m = p$; M_n represents the averaged number of successes in n trials and $M_n \to p$, with probability one, as $n \to \infty$.

If the variation is finite (which is a sufficient but not necessary condition) Chebyshev's inequality, see [14], can be used to prove the weak law of large numbers, namely,

$$\mathbb{P}(|M_n - m| \geq \varepsilon) \leq \frac{\text{Var}(M_n)}{n\varepsilon^2} \to 0$$

as $n \to \infty$.

The central limit theorem specifies the random variable that the (scaled) difference $M_n - m$ converges to. Let $N(0,1)$ be the standard random variable with a normal distribution, means zero, and variance 1.

Theorem 1.2 (The Central Limit Theorem). *Suppose that $\{X_k\}$ is an independent sequence of random variables, with the same distribution, with mean m and a finite positive variances σ^2. Then*

$$\lim_{n\to\infty} \frac{\sqrt{n}(M_n - m)}{\sigma} = N(0,1)$$

in distribution.

For a proof of the central limit theorem and the various ways in which the hypothesis of independence can be weakened and for the following example, see [14].

Example 1.2 (The de Moivre–Laplace Theorem).
Let X_k take the values 1 and 0 with probability p and $1 - p$, respectively. Then $m = p$ and $\sigma^2 = p(1 - p)$. M_n is the averaged number of successes in n trials and

$$\lim_{n\to\infty} \frac{\sqrt{n}(M_n - p)}{\sqrt{p(1-p)}} = N(0,1)$$

in distribution.

The large deviation principle characterizes the limiting behavior, as $\varepsilon \to 0$ (think of ε as $\frac{1}{n}$), of a family of probability measures $\{\mu_\varepsilon\}$ on $(X, \mathscr{B})$ in terms of a rate function. The characterization is in terms of asymptotic upper and lower exponential bounds on the values that μ_ε assigns to measurable subsets of X. Here X is a complete metric space and $\mathscr{B}$ is the Borel σ-field. (Typically X is a function space or a space of measures.)

Definition 1.1. A rate function I is a nonnegative, lower semi-continuous mapping $I : X \to [0, \infty)$, with level sets $\{x : I(x) \leq l\}$ that are closed subsets of X, for each

$l < \infty$. A good rate function is a rate function for which all the level sets are compact subsets of X. The domain of I, $\mathscr{D}_I$ is the set of points of X that have a finite rate, $I(x) < \infty$.

Definition 1.2. $\{\mu_\varepsilon\}$ satisfies the large deviation principle with a rate function I if for all $U \in \mathscr{B}$,

$$- \inf_{x \in U^o} I(x) \leq \liminf_{\varepsilon \to 0} \varepsilon \log(\mu_\varepsilon(U)) \tag{1.38}$$

and

$$\limsup_{\varepsilon \to 0} \varepsilon \log(\mu_\varepsilon(U)) \leq - \inf_{x \in \bar{U}} I(x). \tag{1.39}$$

Here U^o and $\bar{U}$ denote the interior and the closure of the Borel set U. In particular if

$$\inf_{x \in U^o} I(x) = \inf_{x \in U} I(x) = \inf_{x \in \bar{U}} I(x)$$

then

$$\lim_{\varepsilon \to 0} \varepsilon \log P_\varepsilon(U) = - \inf_{x \in U} I(x).$$

1.5.1 Cramér's Theorem

The most common example of large deviation is when μ_n is the distribution on the real line corresponding to the mean

$$M_n = \frac{1}{n} \sum_{j=1}^{n} p_j$$

of independent random variables with a common distribution μ. Suppose that the moment-generating function

$$M(\theta) = E(\exp(\theta p)) = \int_{-\infty}^{\infty} \mathrm{e}^{\theta x} \mathrm{d}\mu(x)$$

is finite for all θ. We assume, for simplicity, that the random variable p is neither bounded above nor below; see [73]. Consider the function

$$I(x) = \sup_{\theta} [\theta x - \log M(\theta)]. \tag{1.40}$$

Then we get the following theorem.

Theorem 1.3 (Cramér's Theorem). *The sequence of probability measures μ_n satisfies the large deviation principle with the good rate function (1.40).*

For proof, see [73].

Example 1.3 (The Intensity Function for Poisson Random Variables). Let $\{p_j\}$ be Poisson distributed with rate λ, then the moment-generating function of a random variable p with distribution μ is

$$M(\theta) = \sum_{n=0}^{\infty} \frac{(\lambda e^{\theta})^n e^{-\lambda}}{n!} = e^{\lambda(e^{\theta}-1)}.$$

Cramér's function is

$$I(x) = \max_{\theta}(x\theta - \ln M(\theta)) = \max_{\theta}(x\theta - \lambda(e^{\theta}-1)) = x\ln\left(\frac{x}{\lambda}\right) - x + \lambda,$$

by differentiation, with respect to θ to find the maximum, and substituting in the maximum.

In higher dimensions Cramér's theorem is similar but one has to work with convex sets. Suppose that μ is a probability distribution $\mathbb{R}^n$, $\theta \in \mathbb{R}^n$ and that the momentum generating function

$$M(\theta) = E(\exp < \theta, p >) = \int_{\mathbb{R}^n} \exp < \theta, x > \mu(dx)$$

is finite for every θ. Now define the rate function

$$I(y) = \sup_{\theta}[< \theta, y > - \log M(\theta)]. \tag{1.41}$$

Then $I(\cdot)$ is convex, nonnegative, semi-continuous (actually continuous) and has the minimum value 0, at the mean $y = m$ of the distribution μ:

$$m = \nabla M(0) = \int_{\mathbb{R}^n} x\mu(dx).$$

Let $\{\mu_k\}$ be as above and the distributions of $\{M_k\}$ have a common distribution μ, then the large deviation principle holds.

Theorem 1.4. *The sequence of probability measures μ_k satisfies the large deviation principle with the good rate function (1.42).*

The main new fact needed in the proof is the minimax theorem, which states that for any compact convex set $U \subset \mathbb{R}^n$

$$\inf_{y\in U} I(y) = \inf_{y\in U}\sup_{\theta}[< \theta, y > - \log M(\theta)] = \sup_{\theta}\inf_{y\in U}[< \theta, y > - \log M(\theta)]; \tag{1.42}$$

see [73].

1.5.2 Stochastic Processes and Time Change

We follow [12] in defining a stochastic process.

Definition 1.3. Given an index set I, a stochastic process indexed by I is a collection of random variables $\{x_\lambda\}$ on a probability space $(\Omega, \mathscr{F}, \mathbb{P})$ taking values in a set S. The set S is called the state space of the process.

We will take $I = \mathbb{R}^+$ in most of this book and $S = \mathbb{R}^n$ or the natural numbers $\mathbb{N}$ (including zero). In this case the stochastic process will be denoted respectively x_t, or N_t, $t \geq 0$. Many more examples and properties of stochastic processes can be found in [12]. Our main examples in Sect. 1.6 below will be the Poisson process and the Brownian motion process.

We now consider the large deviation principle with a time change. Suppose that x_t is a stochastic process with stationary independent increments that has no Gaussian component. Suppose that the infinitely divisible random variable x_T has a finite moment-generating function $M(\theta)$ for θ bounded. Assume that x_T has no Gaussian component. Let μ be a probability measure on $[0,T]$ and define

$$z_t^\varepsilon = \varepsilon x(\varepsilon^{-1}\mu[0,t]), \quad \text{for } 0 \leq t \leq T.$$

Let $f \in BV[0,T]$ be a function of bounded variations on $[0,T]$, then $f = f_1 + f_2$, where $f_1 \ll \mu$ and $f_2 \perp \mu$. Moreover, f_2 has a Hahn decomposition:

$$f_2 = h_1 - h_2,$$

where $h_1, h_2 \in M[0,T]$ are bounded nonnegative measures on $[0,T]$. The following theorem holds; see [43].

Theorem 1.5. *Let $\{\mu_\varepsilon\}$ be the probability distributions of $z_\varepsilon(t)$, $0 \leq t \leq T$, then $\{\mu_\varepsilon\}$ satisfies the large deviation principle with the rate function*

$$I(f) = \int_{\mathbb{R}} I(\dot{f})\mathrm{d}\mu + C_1 h_1([0,T]) + C_2 h_2([0,T]), \tag{1.43}$$

where $f \in BV[0,T]$, $f = h_1 - h_2$, and

$$C_1 = \lim_{x\to\infty} \frac{I(x)}{x}, \; C_2 = \lim_{x\to-\infty} \frac{I(x)}{x}.$$

The proof can be found in [43].

Remark 1.1. In Sect. 1.7 we will use the central limit theorem and the large deviation principle to construct the noise in fully developed turbulence. It is neither necessary that the dissipation processes are independent nor Poisson processes. The assumption of independence can be weakened in various ways in the central limit theorem, as mentioned above, and the deterministic bounds that we get from the large deviation principle hold for a much larger class of processes than just Poisson processes. The way to think about this is that the central limit theorem captures the mean of the dissipation, and the large deviation principle gives a bound on the fluctuations (large excursion) in the mean of the dissipation, which can have non-trivial correlations. Then we will also need a multiplicative noise term to capture the large excursions of the velocity, but this is all that is required to obtain a universal stochastic forcing.

Example 1.4 (Poisson Processes). Let $\{x_t^j\}$ be Poisson processes with a constant rate λ. Consider their average M_n and let $z_t^{\{\frac{1}{n}\}} = \frac{1}{n}(M_n - m)(n\mu[0,t])$ as above. Here μ is Lebesgue measure. Notice that the scaling $\frac{1}{n}$ is different from the scaling $\sqrt{n}$ above that gives the Brownian motion. Then an application of Theorem 1.5 gives that $z_t^{\{\frac{1}{n}\}}$ satisfies the large deviation principle with the rate function

$$I(f) = \int_{\mathbb{R}} \dot{f} \log\left(\frac{\dot{f}}{\lambda}\right) \mathrm{d}\mu + \lambda - f[0,T], \tag{1.44}$$

if f is absolutely continuous with respect to Lebesgue measure, and $I = \infty$, otherwise. Here $\dot{f} = \frac{\mathrm{d}f}{\mathrm{d}t}$, $f = f_1$, $h_1 = h_2 = 0$, and $C_1 = \infty$.

1.6 Poisson Processes and Brownian Motion

The Poisson process is the simplest random process imaginable when the random variables are independent. If we start with the binomial distribution investigated by S.-D. Poisson in 1837, it gives the probability of r heads and $n-r$ tails in n tosses of a coin, with probability p getting a head and probability $1-p$ getting a tail:

$$b(n,p;r) = \binom{n}{r} p^r (1-p)^{n-r}.$$

$\binom{n}{r} = \frac{n!}{r!(n-r)!}$ is the binomial coefficient. If we keep the average number of heads $\lambda = np$ constant, while $n \to \infty$ and $p \to 0$, then we get a limit

$$\pi_r(\lambda) = \lim_{n\to\infty} b(n,p;r) = \lambda^r \frac{\mathrm{e}^{-\lambda}}{r!}$$

for $r \geq 0$. This limit is called the Poisson distribution $\mathscr{P}(\lambda)$ with parameter (rate) λ.

In the context of turbulence we will be interested in how often a dissipation event or a velocity excursion takes place in a set $A \subset \mathbb{R}^3$ in three-dimensional space. Our random variable counts the number of these events in A. in an interval of time.

A Poisson random variable X has the Poisson distribution where the possible values of X are nonnegative integers and

$$\mathbb{P}(X = n) = \pi_n(\lambda) = \lambda^n \frac{\mathrm{e}^{-\lambda}}{n!} \tag{1.45}$$

for $n \geq 0$; see [33]. We will denote the distribution concentrated at 0 by $\mathscr{P}(0)$:

$$\mathbb{P}(X = 0) = 1$$

and the distribution concentrated at ∞ by $\mathscr{P}(\infty)$:

$$\mathbb{P}(X = \infty) = 1.$$

It immediately follows that the mean is

$$E(X) = \sum_{n=0}^{\infty} n\lambda^n \frac{\mathrm{e}^{-\lambda}}{n!} = \lambda \tag{1.46}$$

and if z is a complex number, then z^X is finite so that

$$E(z^X) = \sum_{n=0}^{\infty} z^n \lambda^n \frac{\mathrm{e}^{-\lambda}}{n!} = \mathrm{e}^{-\lambda} \sum_{n=0}^{\infty} \frac{(\lambda z)^n}{n!} = \mathrm{e}^{-\lambda(1-z)}. \tag{1.47}$$

The moments follow by differentiation:

$$\begin{aligned} E(X) &= \lambda \\ E(X(X-1)) &= \lambda^2 \\ E(X(X-1)(X-2)) &= \lambda^3, \end{aligned}$$

etc. This gives the moments

$$E(X) = \lambda,\ \mathrm{Var}(X) = \lambda + \lambda^2,\ E(X^3) = \lambda + 3\lambda^2 + \lambda^3, \tag{1.48}$$

etc.

A great simplifying property of Poisson random variables is their additivity or the closure of the distribution under convolution. Let X and Y be independent random variables with Poisson distributions $\mathscr{P}(\lambda)$ and $\mathscr{P}(\mu)$, respectively. Then for $r \geq 0,\ s \geq 0$,

$$\mathbb{P}(X = r,\ Y = s) = \mathbb{P}(X = r)\mathbb{P}(Y = s) = \lambda^r \frac{\mathrm{e}^{-\lambda}}{r!} \mu^s \frac{\mathrm{e}^{-\mu}}{s!}.$$

If we sum X and Y then the distribution of $X + Y$ is

$$\begin{aligned} \mathbb{P}(X+Y=n) = \mathbb{P}(X=r,\ Y=n-r) &= \sum_{r=0}^{n} \lambda^r \frac{\mathrm{e}^{-\lambda}}{r!} \mu^{n-r} \frac{\mathrm{e}^{-\mu}}{(n-r)!} \\ &= \frac{\mathrm{e}^{-(\lambda+\mu)}}{n!} \sum_{r=0}^{n} \binom{n}{r} \lambda^r \mu^{n-r} = \frac{(\lambda+\mu)^n \mathrm{e}^{-(\lambda+\mu)}}{n!}. \end{aligned}$$

This shows that $X + Y$ has the distribution $\mathscr{P}(\lambda + \mu)$. In fact this holds for a countable number of random variables,

Theorem 1.6. *Let $X_j,\ j = 1, 2, \ldots$ be independent random variables and assume that X_j has the Poisson distribution $\mathscr{P}(\lambda_j)$. If*

$$\lambda = \sum_{j=1}^{\infty} \lambda_j \tag{1.49}$$

converges, then

$$X = \sum_{j=1}^{\infty} X_j \tag{1.50}$$

converges with probability 1 and has the distribution $\mathscr{P}(\lambda)$. *If on the other hand (1.49) diverges, then X diverges with probability 1.*

See [33] for the statement and the proof of the theorem.

We will now let $(\Omega, \mathscr{F}, \mathbb{P})$ be a probability space and define the Poisson processes. A Poisson process is a map Π from Ω into a state space S, for example, $\mathbb{R}^3$, such that the count function

$$N(A) = \#\{\Pi \cap A\}$$

is a well-defined random variable. In other words Π lies in the countable subsets S^∞ of the state space S, for all measurable sets $A \subset S$.

Suppose that S is a measurable space such that the diagonal

$$D = \{(x,y); x = y\}$$

is measurable on the product space $S \times S$. This implies that every singleton $\{x\} \in S$ is measurable.

Definition 1.4. A Poisson process on S is a random countable subset Π of S such that:

1. For any disjoint measurable subsets $A_1, A_2, \dots, A_n$ of S, the random variables $N(A_1), N(A_2), \dots, N(A_n)$ are independent.
2. $N(A)$ has the Poisson distribution $\mathscr{P}(\lambda)$ where $0 \le \lambda \le \infty$.
3. The measure

   $$\mu(A) = E(N(A))$$

   is the mean measure (or the Lévy measure) of the Poisson process Π.

This definition implies that if $\mu(A)$ is finite, then A is finite with probability 1 and empty if $\mu(A) = 0$. If $\mu(A) = \infty$, $\Pi \cap A$ is countably infinite with probability 1. It also follows that if $A = \bigcup_{j=1}^{\infty} A_j$ and the A_js have empty intersections, then

$$N(A) = \sum_{j=1}^{\infty} N(A_j)$$

and

$$\mu(A) = \sum_{j=1}^{\infty} \mu(A_j),$$

see [33] for more properties and information.

We now define Poisson processes with a continuous parameter t following [12].

Definition 1.5. The Poisson process $\{N_t; t \geq 0\}$ with intensity function ρ is a process with state space $S = \{0,1,2,\ldots\} = \mathbb{N}$, having independent increments distributed as

$$\mathbb{P}(N_t - N_s = j) = \frac{(\int_s^t \rho(r)\mathrm{d}r)^j}{j!} \exp\left(-\int_s^t \rho(r)\mathrm{d}r\right)$$

for $j = 0,1,2,\ldots$, $s < t$, where $\rho(r), r \geq 0$, is a continuous nonnegative function.

Because of the independent increments N_t is a Markov process; see [12]. In the case $\rho(r) = \lambda$ (constant), the transition probabilities become

$$\begin{aligned} p_{ij}(s,t) &= \frac{[\lambda(t-s)]^{j-s}}{(j-s)!} \mathrm{e}^{-\lambda(t-s)}, \quad i \leq j \\ &= 0, \quad j > i, \end{aligned}$$

so $p_{ij}(s,t) = p_{ij}(t-s)$ and the transition law is time-homogeneous. In this case the process is referred to as the Poisson process with parameter (or rate) λ.

Definition 1.6 (The Compound Poisson Process). Let $\{N_t\}$ be a Poisson process with parameter $\lambda > 0$ starting at 0, and let $Y_1, Y_2, \ldots$ be independent, identically distributed random variables, independent of $\{N_t\}$, and having a common probability mass function f. The process $\{x_t\}$ defined by

$$x_t = \sum_{k=0}^{N_t} Y_k,$$

where Y_0 is independent of $\{N_t\}$ and the processes $Y_1, Y_2, \ldots$, is called the compound Poisson process.

The compound Poisson process also has independent increments and is a Markov process. Its transition probabilities are given by

$$\begin{aligned} p_{ij}(s,t) &= E(\mathbb{P}(x_t - x_s = j - i | N_t - N_s)) \\ &= \sum_{k=0}^{\infty} f^{*^k}(j-i) \frac{[\lambda(t-s)]^k}{k!} \mathrm{e}^{-\lambda(t-s)}, \end{aligned}$$

where f^{*^k} is the kth-fold convolution of f with $f^{*^0}(0) = 0$. Thus $\{x_t\}$ also has a time-homogeneous transition law; see [12] for more information.

Example 1.5 (Log-Poisson Processes).

1. Suppose the Poisson process N_k has the mean $\lambda = -\frac{\gamma \ln|k|}{\beta - 1}$, then it is straightforward to compute the mean of the log-Poisson process $|k|^\gamma \beta^{N_k}$. Namely,

$$E(|k|^\gamma \beta^{N_k}) = \sum_{j=0}^{\infty} |k|^\gamma \beta^j \frac{\lambda^j}{j!} \mathrm{e}^{-\lambda} = |k|^\gamma \sum_{j=0}^{\infty} \frac{(\beta\lambda)^j}{j!} \mathrm{e}^{-\lambda} = |k|^\gamma \mathrm{e}^{(\beta-1)\lambda}.$$

Thus

$$\ln[E(|k|^{\gamma}\beta^{N_k})] = \gamma \ln|k| + (\beta-1)\lambda = \gamma \ln|k| - (\beta-1)\frac{\gamma}{\beta-1}\ln|k| = 0,$$

and we get the mean $E(|k|^{\gamma}\beta^{N_k}) = 1$.

2. Now we compute the $p/3$ moment $E([|k|^{\gamma}\beta^{N_k}]^{p/3})$ of the log-Poisson process $|k|^{\gamma}\beta^{N_k}$ above. By a similar computations as above,

$$E([|k|^{\gamma}\beta^{N_k}]^{p/3}) = |k|^{\frac{p\gamma}{3}} e^{(\beta^{p/3}-1)\lambda},$$

where λ is the mean from part 1; therefore,

$$\ln[E([|k|^{\gamma}\beta^{N_k}]^{p/3})] = \frac{p}{3}\gamma \ln|k| + (\beta^{p/3}-1)\lambda = \left(\frac{p}{3} - \frac{(\beta^{p/3}-1)}{\beta-1}\right)\gamma \ln|k|.$$

Finally, we get that

$$E([|k|^{\gamma}\beta^{N_k}]^{p/3}) = |k|^{\gamma\left(\frac{p}{3} - \frac{(\beta^{p/3}-1)}{\beta-1}\right)}.$$

1.6.1 Finite-Dimensional Brownian Motion

The Scottish botanist Robert Brown observed in 1828 that grains of pollen in liquid execute an irregular motion, which was later explained as the result of many random collisions with the molecules of the liquid. It is natural to model the motion of the grains ω by a stochastic process $b_t(\omega)$ interpreted as the position at time t of the grain. We will consider the motion of such a process in $\mathbb{R}^n$; see [51].

We first let $x \in \mathbb{R}^n$ and define the Gaussian (or normal) probability distribution

$$p(t,x,y) = \frac{1}{(2\pi t)^{n/2}} \exp\left(-\frac{|x-y|^2}{2t}\right), \quad y \in \mathbb{R}^n, \ t \geq 0. \tag{1.51}$$

Before defining Brownian motion we review the basic properties of normal random variable following Oksendal [51].

Definition 1.7. Let $(\Omega, \mathscr{F}, \mathbb{P}^x)$ be a probability space. A random variables $X : \Omega \to \mathbb{R}$ is normal if it has the PDF,

$$p_X(x) = \frac{1}{\sigma\sqrt{2\pi}} \exp\left(-\frac{(x-m)^2}{2\sigma^2}\right), \tag{1.52}$$

where $\sigma \geq 0$ and m are constants. In other words,

$$\mathbb{P}[X \in A] = \int_A p_X(x) \mathrm{d}x$$

for all Borel sets $A \subset \mathbb{R}$.

Then the mean of X is

$$E(X) = \int_{\Omega} X \mathrm{d}\mathbb{P} = \int_{\mathbb{R}} x p_X(x) \mathrm{d}x = m \tag{1.53}$$

and the variance of X is

$$\mathrm{var}(X) = E([X-m]^2) = \int_{\mathbb{R}} (x-m)^2 p_X(x) \mathrm{d}x = \sigma^2. \tag{1.54}$$

More generally $X : \Omega \to \mathbb{R}^n$ is a normal variable in n dimensions, with a law $\mathcal{N}(m,C)$, if the distributions of X have the density

$$p_X(x) = \frac{\sqrt{|C^{-1}|}}{\sqrt{(2\pi)^n}} \exp(-\langle (x-m), C^{-1}(x-m) \rangle), \tag{1.55}$$

where the mean of X is

$$E(x) = \int_{\Omega} X \mathrm{d}\mathbb{P} = \int_{\mathbb{R}} x p_X(x) \mathrm{d}x = m \tag{1.56}$$

and C^{-1} is a symmetric positive definite matrix that is the inverse of the covariance matrix of X,

$$E(\langle (X-m), (X-m) \rangle) = C = [c_{j,k}] = [E(\langle (X_j - m_j), (X_k - m_k) \rangle)] \tag{1.57}$$

x and m are n vectors in $\mathbb{R}^n$ and $C \in \mathbb{R}^{n \times n}$ and $\langle \cdot, \cdot \rangle$ denotes the (inner) product of vectors in $\mathbb{R}^n$.

Definition 1.8. The characteristic function of a random variable $X : \Omega \to \mathbb{R}^n$ is the Fourier transform of X, $\phi_X(h) : \mathbb{R}^n \to \mathbb{C}$:

$$\phi_X(h) = E(\exp(\mathrm{i}\langle X, h \rangle)) = \int_{\mathbb{R}^n} \mathrm{e}^{\mathrm{i}\langle X, h \rangle} \mathbb{P}(X \in \mathrm{d}x), \tag{1.58}$$

$h \in \mathbb{R}^n$.

It is now clear that the characteristic function determines the random variable $X \in L^2(\Omega)$ uniquely, since it is the Fourier transform, and for a normal variable the characteristic function is

$$\phi_X(h) = \exp\left(-\frac{1}{2}\langle h, Ch \rangle + \mathrm{i}\langle h, m \rangle\right). \tag{1.59}$$

Using the characteristic function it is now easy to show that $X = (X_1, \ldots, X_n)$ is constituted of normal variables if and only if any linear combination $Y = \sum_{j=1}^{n} X_j$ is normal, see [51] and that real normal variables X_0 and $X_1, \ldots, X_n$ are independent if and only if $E(\langle (X_0 - m_0), (X_j - m_j) \rangle) = 0$ for all $j = 1, \ldots, n$. We end with a theorem from Oksendal [51] about the convergence of normal variables.

Theorem 1.7. *Suppose that $X_k : \Omega \to \mathbb{R}^n$ are normal for all k and that $X_k \to X$ as $k \to \infty$, or*

$$E(|X_k - X|^2) \to 0, \text{ as } k \to \infty.$$

Then X is normal.

Proof. Since

$$|e^{i\langle h,x\rangle} - e^{i\langle h,y\rangle}| \le |u| \cdot |x-y|,$$

$$E([e^{i\langle h,X_k\rangle} - e^{i\langle h,X\rangle}]^2) \le |h|^2 \cdot E(|X_k - X|^2) \to 0$$

as $k \to \infty$. Thus

$$E(e^{i\langle h,X_k\rangle}) \to E(e^{i\langle h,X\rangle})$$

as $k \to \infty$. By use of the characteristic function above, this implies that X is normal with mean $E(X) = \lim_{k\to\infty} E(X_k)$ and covariance $C = \lim_{k\to\infty}$ covariance(X_k).

1.6.2 The Wiener Process

We are now ready to define Brownian motion or a Wiener process following [13, 14]. This process will be infinite-dimensional in the sense that the paths $b_t(\omega)$ will be functions lying in the infinite-dimensional function space $C[0,\infty)$. However, for t_0 fixed, we still have $b_{t_0} \in \mathbb{R}$. If $b_t \in \mathbb{R}^n$, then each component of $b_t(\omega)$ lies in $C[0,\infty)$ and below we will let $0 \in \mathbb{R}^n$ denote an n-vector with zero entries.

Definition 1.9. A Brownian motion or a Wiener process is a stochastic process $\{b_t; t \ge 0\}$ on some probability space $(\Omega, \mathscr{F}, \mathbb{P})$, having the following four properties:

1. The process starts at 0:

$$\mathbb{P}(b_0 = 0) = 1.$$

2. The increments are independent: If $0 \le t_0 \le t_1 \le \cdots \le t_k$ and $A_i \in \mathscr{F}$, then

$$\mathbb{P}(b_{t_i} - b_{t_{i-1}} \in A_i, i \le k) = \prod_{i \le k} \mathbb{P}(b_{t_i} - b_{t_{i-1}} \in A_i).$$

3. For $s \le 0 \le t$ the increments $b_t - b_s$ are normally distributed with mean 0 and variance $n(t-s)$:

$$\mathbb{P}(b_t - b_s \in A) = \frac{1}{(2\pi(t-s))^{n/2}} \int_A e^{-\frac{|x|^2}{2(t-s)}} dx.$$

4. For each $\omega \in \Omega$, the path $b_t(\omega)$ is continuous in t and $b_0(\omega) = 0$.

The distribution $\mathbb{P} \circ b^{-1}$ of the process $b = \{b_t; t \ge 0\}$ is a probability measure concentrated on the Borel σ-field of $C[0,\infty)$, referred to as Wiener measure.

Now the finite-dimensional distributions of b_t are given by

$$\mathbb{P}(b_{t_1} \in A_1, b_{t_2} \in A_2, \ldots, b_{t_k} \in A_k) \quad (1.60)$$
$$= \int_{A_1 \times \cdots \times A_k} p(t_1, x, x_1) \cdots p(t_k - t_{k-1}, x_{k-1}, x_k) \mathrm{d}x_1 \cdot \mathrm{d}x_k,$$

where $A_j \in \mathscr{F}$. The existence of $(\Omega, \mathscr{F}, \mathbb{P})$ and b_t is given by Kolmogorov's extension theorem; see [14, 51]. However, Kolmogorov's theorem does not imply Condition 4. The problem is that the Kolmogorov σ-field does not include the set of continuous functions $C[0,\infty)$. It only consists of sets determined by countably many coordinates and this is insufficient to determine whether a function is continuous, a question involving uncountably many points; see [12] for more information on this issue. Consequently, we have to use the construction by Wiener of Brownian motion; see [14]. A simpler modern construction, the Lévy–Ciesielski construction by wavelets, is given in [13]. We refer the reader to these references for the proof of Condition 4. Together Kolmogorov's theorem and Wiener's construction give the following theorem; see [14].

Theorem 1.8. *There exists a stochastic process called Brownian motion, satisfying Conditions 1–4 in Definition 1.9.*

We can now show that Brownian motion has the following properties, see [51]:

1. Brownian motion is normal (a Gaussian process) with law $\mathscr{N}(m, \sigma^2)$. The mean of a one-dimensional Brownian motion starting at $b_0 = m$ is (we will omit the dependence of E on m, $E = \mathrm{e}^m$),

   $$E(b_t) = E(b_t - m + m) = \int_{\mathbb{R}} [(x - m) + m] p_X(x) \mathrm{d}x = m$$

   by (1.53). Similarly, the variance is

   $$E([b_t - m]^2) = \sigma^2 = t$$

   by (1.53) and (1.51) where $\sigma^2 = t$. Now if $b_t \in \mathbb{R}^n$ is an n vector, then its mean is also an n vector:

   $$E(b_t) = m \in \mathbb{R}^n, \quad (1.61)$$

   and the variance is

   $$E([b_t - m]^2) = n\sigma^2 = nt, \quad (1.62)$$

   where we have summed the contributions t from the n entries of b_t. It is easy to see that the characteristic function of b_t evaluated at k times, $0 \le t_1 \le t_2 \cdots \le t_k$, is of the form (1.59), see [51], and therefore b_t is normal. Then one computes

   $$E([b_t - m][b_s - m]) = n \min(t, s) \quad (1.63)$$

   and

   $$E([b_t - b_s]^2) = n(t - s). \quad (1.64)$$

Namely, (1.63) follows from (1.62) and (1.64) by the computation

$$\begin{aligned} E([b_t - b_s]^2) &= E([b_t - m]^2 - 2[b_t - m][b_s - m] + [b_s - m]^2) \\ &= n(t - 2s + s) = n(t - s), \end{aligned}$$

if $t \geq s$; see [51].

2. Brownian motion has independent increments. Namely, the random variables

$$b_{t_1}, b_{t_2} - b_{t_1}, \ldots, b_{t_k} - b_{t_{k-1}}$$

are independent for all $0 \leq t_1 \leq t_2 \cdots \leq t_k$. We prove this by showing that these variables are uncorrelated. It implies the they are independents as discussed above:

$$E([b_{t_i} - b_{t_{i-1}}][b_{t_j} - b_{t_{j-1}}]) = 0, \text{ for } t_i \leq t_{j-1}.$$

This follows from (1.63):

$$\begin{aligned} E([b_{t_i} - b_{t_{i-1}}][b_{t_j} - b_{t_{j-1}}]) &= E(b_{t_i} b_{t_j} - b_{t_{i-1}} b_{t_j} - b_{t_i} b_{t_{j-1}} + b_{t_{i-1}} b_{t_{j-1}}) \\ &= n(t_i - t_{i-1} - t_i + t_{i-1}) = 0. \end{aligned}$$

3. Brownian motion has continuous paths. This follows from Wiener's construction; see [13].

We leave it to the reader to compute the Fourier transform of a Gaussian,

$$\int_{\mathbb{R}} \frac{e^{ih(x-m)}}{\sqrt{2\pi}\sigma} e^{-\frac{(x-m)^2}{2\sigma^2}} dx = e^{-\frac{1}{2}h^2 + i\, hm}$$

and use it to prove (1.59). The characteristic function $E(e^{ib_t h}) = e^{-\frac{1}{2}h^2}$ can also be used to compute the moments

$$E(b_t^{2k}) = \frac{(2k)!}{2^k k!} t^k, \ k \in \mathbb{N},$$

and $E(b_t^{2k+1}) = 0$, of Brownian motion, and to prove that $b_t - b_{t_0}$, with t_0 fixed, is also a Brownian motion.

The compound Poisson process and Brownian motion are the building blocks of all processes with independent increments; see [12, 13]. We will see in the next section how the generic noise in turbulence can be expressed in terms of these processes.

1.7 The Noise in Fully Developed Turbulence

We will assume that the fluid satisfies periodic boundary conditions on its domain. This is done for convenience and can easily be relaxed. Then the velocity lies in a nice Hilbert space, namely, $u(x) \in L^2(\mathbb{T}^3)$, or the underlying domain $\mathscr{D}$ can be taken to be a three torus $\mathbb{T}^3$, and the fluid velocity lies in the space of functions square integrable on the torus. By a classical result by Leray [42], see Sect. 4.1, one knows that if $\nabla u(x,0)$ lies in L^2, then $u(x,t)$ lies in L^2, for all t, and that one can also make sense of the gradient ∇u for almost every t, at least for the deterministic equation (1.1).

The stochastic Navier–Stokes equation describing fully developed turbulence is

$$\begin{aligned} du &= (\nu\Delta u - u\cdot\nabla u + \nabla\Delta^{-1}\mathrm{tr}(\nabla u)^2)dt + \sum_{k\in\mathbb{Z}^3} c_k^{\frac{1}{2}}\,db_t^k e_k(x) \\ &\quad + \sum_{k\neq 0} d_k\eta_k dt e_k(x) + u\sum_{k\neq 0}^{m}\int_{\mathbb{R}} h_k\bar{N}^k(dt,dz) \qquad (1.65) \\ u(x,0) &= u_0(x), \end{aligned}$$

where, in the additive noise, each Fourier component $e_k = e^{2\pi i k\cdot x}$ comes with its own independent Brownian motion b_t^k and a deterministic term $\eta_k t$. The coefficients $c_k^{\frac{1}{2}}$ and d_k decay sufficiently fast so that the Fourier series converges. The sizes of the jumps h_k in the velocity gradient do not decay, but for $t<\infty$, only finitely many $h_k s$, $|k|\leq m$, are nonzero.

The stochastic processes b_t^k are independent. The discrete processes N_t^k are also independent, for different ks, but can be associated with b_k and $\eta_k t$, for the same k. This link is manifested in the experimentally observed fact that large velocity excursion is accompanied by large dissipation events.

The situation described by (1.65) is the general situation in turbulent flow. There is some large-scale flow that drives all the small scale and one can decompose the velocity field into two parts $U+u$ where U describes the large-scale flow and u describes the smaller-scale turbulence. In physics u is said to describe the fluctuations. The large-scale flow generates a force acting on the small scale and the noise in (1.65) is a model of this force. We will argue below that based on probability theory this force has a general form in fully developed turbulence. This decomposition of the velocity field can also be thought of as the classical Reynolds decomposition and then the force, exerted by the small scales u on the large scales U, is the well-known eddy viscosity. Still another way of thinking about (1.65) is in terms of the coarse graining of the Navier–Stokes equation, where U describes the mean flow and (1.65) is the equation describing the fluctuations u.

Turbulent flow consists of complicated and sometimes violent motion that is dissipated in the flow. We split the torus into small boxes and let p_j denote the stochastic dissipation process in the jth box. We assume that the p_js in different boxes are weakly coupled and have mean m. By the Central Limit Theorem 1.2, in probability theory, the average

$$M_n = \frac{1}{n}\sum_{j=1}^{n} p_j$$

converges to a normal (Gaussian) random variable $\sqrt{n}(M_n - m)/\sigma \to N(0,1)$ as $n \to \infty$, with mean zero and variance one, as we let the number of boxes (n) increase to infinity. We now let

$$S_n = \sum_{j=1}^{n} p_j$$

denote the sum and define the stochastic processes:

$$x_t^n = \frac{S_{[tn]} - nm}{\sqrt{n}\sigma},$$

where $[tn]$ denotes integer value. Then if the p_js are independent and identically distributed with variance $\sigma^2 > 0$ and mean m, the Functional Central Limit Theorem, see Theorem 8.1 in [12], says that the stochastic processes $\{x_t^n, t \geq 0\}$ converge (in distribution) to a Brownian motion b_t, starting at the origin with zero drift and diffusion coefficient 1, as $n \to \infty$. This must be done in the direction of any Fourier component ($e_k = \exp(2\pi i k \cdot x)$), that forms a basis in the infinite-dimensional space L^2, and the result is the differential of an infinite-dimensional Brownian motion

$$\mathrm{d}f_t^1 = \sum_{k \in \mathbb{Z}^3} c_k^{\frac{1}{2}} \mathrm{d}b_t^k e_k(x).$$

Here each Fourier component comes with its independent Brownian motion b_t^k and the $c_k^{1/2}$s are constant vectors.

The Central Limit Theorem 1.2 says that the average of the dissipation processes converges to a Gaussian, but there also exist large excursion or fluctuations in the mean. The effects of these fluctuations are frequently captured by the large deviation principle; see Definition 1.2. If these excursions are completely random, then they can, for example, be modeled by a Poisson process with the rate λ. If, moreover, these processes have a bias, an application of the large deviation principle, Definition 1.2, shows that the large deviations of M_n are bounded above by a deterministic term which is a constant determining the direction of the bias, times the rate η. By Theorems 1.3 and 1.5 and Examples 1.3 and 1.4, since the rate $\lambda_k \to \infty$ as $k \to \infty$, the rate function is bounded by $\eta = \lambda$. This also holds in the direction of each Fourier component and gives the term,

$$\mathrm{d}f_t^2 = \sum_{k \neq 0} \mathrm{d}_k \eta_k \mathrm{d}t e_k(x),$$

the second term in the additive noise in stochastic Navier–Stokes equation. Here the d_ks are constant vectors, representing the bias in a particular direction in Fourier space, and the η_k are the rates in the kth direction. We will choose the rate $\eta_k = |k|^{1/3}$ below. This makes the two terms in the additive noise give similar scaling in the Fourier variable k. This must be the case, because the second term in capturing the fluctuations in the mean by an application of the Large Deviation Principle 1.2, and thus together the two terms give a more accurate description of the mean. In other words there is only one additive noise term $df_1 + df_2$. It turns out, see below, that together the two terms produce the Kolmogorov–Obukhov '42 scaling. Intermittency in the dissipation is then an additional effect caused by the interaction of the multiplicative and additive noise with the Navier–Stokes evolution. This will be made clear below.

We must also capture the large excursions and intermittency in the velocity and this gives rise to a multiplicative noise term (multiplying the velocity) in the stochastic Navier–Stokes equations. The velocity fluctuations, are discrete and if they are completely random, they can be modeled by the Poisson jump process x_t^k, with its number process N_t^k denoting the integer number of velocity excursions, associated with kth wave number, which has occurred at time t. The differential $dN^k(t) = N^k(t+dt) - N^k(t)$ denotes the number of these excursions in the time interval $(t, t+dt]$. The process

$$\sum_{k\neq 0} \int_{\mathbb{R}} h_k(t,z)\bar{N}^k(dt,dz),$$

in the multiplicative noise, models the excursions (jumps) in the velocity gradient; see [52]. The h_k are the sizes of the jumps in the velocity gradients and $\bar{N}^k$ is the compensated number (of jumps) process. We will include a term in the Poissonian distribution for the jump process that correlates N^k with only the kth Fourier mode. This models the link between large velocity and dissipation events.

Equation (1.65) represents the stochastic Navier–Stokes equation for the small scales with the general form of turbulent noise. The two terms in the additive noise result from scaling the average of the dissipation processes in different ways in n (number of processes), but they must both be present, and together they accurately describe the mean dissipation. The coefficients $c_k^{1/2}$ and d_k give their relative size that varies from experiments to experiment, for small k. For large k this ratio should be universal. The central limit theorem and the large deviation principle determine the additive noise in fully developed turbulence, but the multiplicative noise is modeled in (1.65) as a general (Poisson) jump process. It would also be possible to formulate the equation as the deterministic equation (1.1) if we continuously modified the initial data so as to absorb the evolving noise. This amounts to continuously modifying the initial data with a stochastic process and is what is effectively done in direct Navier–Stokes simulations (DNS). Clearly, these two formulations must be equivalent.

1.7.1 The Generic Noise

In this section we will ask the question: In what sense is the noise in the stochastic Navier–Stokes equation (1.65) generic? The mathematical answer is that it is modeled by a homogeneous Lévy process, which is as general as you would expect the noise in fully developed turbulence to be. A homogeneous Lévy process can be written as a sum of a Brownian motion and a limit of independent superpositions of compound Poisson processes with varying jump sizes; see Theorem 1.3. in [12]. We have used the large deviation principle, Definition 1.2, to estimate probabilities of the Poisson processes, but apart from that, our noise is a perfectly general homogeneous Lévy process. The time-homogeneity means that the increments of your noise process only depend on the time interval that has passed, not on the starting time. This is what one would expect from the noise in fully developed turbulence. In this sense we have generic noise in every Fourier component.

The question still remains why we are representing the noise in (1.65) by a convergent Fourier series? Why do we not take noise that is white both in space and time? Surely, the tiny ambient noise in nature is, white both in space and time. The reason for this is as explained in Sect. 1.4, that we are modeling the noise in fully developed turbulence, not small ambient noise in nature. The latter noise is the source for the noise in fully developed turbulence, but that noise has developed through the Navier–Stokes evolution or the fluid flow, where the tiny white noise gets magnified by the flow instabilities and saturated and colored as explained in Sect. 1.4.

Physically it is also clear that the noise in fully developed turbulence cannot be white both in time and space. The heat equation with such noise is solved in Walsh [75] and found to have continuous solutions only in one or two dimensions. In dimensions three and greater the solutions are distributions without any spatial smoothness. This is contrary to what is observed in turbulent flow. This has direct relevance for the Navier–Stokes equation because the linear part of the equation is the same as that of the heat equation. The fluid velocity seems to be continuous in space even in very-high Reynolds number flow; see Sect. 4.2 for more information on this. The convergence of the Fourier series in (1.65) is the minimal requirement that one can make to get a spatially smoothness of the fluid velocity observed in turbulent flow. In this sense the noise is also physically generic.

1.8 The Stochastic Navier–Stokes Equation for Fully Developed Turbulence

Adding the additive noise given by the Central Limit Theorem 1.2 and the large deviation principle, Definition 1.2, and the multiplicative noise produced by the jumps in velocity to the deterministic equation (1.1), see Sect. 1.7, we get the stochastic Navier–Stokes equations describing fully developed turbulence:

$$\begin{aligned} \mathrm{d}u &= (\nu\Delta u - u\cdot\nabla u + \nabla\Delta^{-1}\mathrm{tr}(\nabla u)^2)\mathrm{d}t + \sum_{k\in\mathbb{Z}^3} c_k^{\frac{1}{2}}\mathrm{d}b_t^k e_k(x) \\ &\quad + \sum_{k\neq 0} \mathrm{d}_k |k|^{1/3}\mathrm{d}t e_k(x) + u\sum_{k\neq 0}^{m}\int_{\mathbb{R}} h_k \bar{N}^k(\mathrm{d}t,\mathrm{d}z), \qquad (1.66) \\ u(x,0) &= u_0(x), \end{aligned}$$

where, in the additive noise, each Fourier component e_k comes with its own independent Brownian motion b_t^k and a deterministic term $|k|^{1/3}t$. The coefficients $c_k^{\frac{1}{2}}$ and d_k decay sufficiently fast so that the Fourier series converges. The sizes of the jumps h_k in the velocity gradient do not decay, but for $t < \infty$, only finitely many $h_k s$, $k \leq m$, are nonzero.

We will now state the basic existence theorem of nonlinear stochastic partial differential equations (SPDEs) in infinite-dimensional space following Da Prato and Zabczyk [57], Theorem 7.4 on page 186. Consider the initial value problem for the SPDE:

$$du = (Au + F(t,u))dt + G(t,u)dB_t, \qquad u(x,0) = u_0. \tag{1.67}$$

We assume that a probability space $(\Omega, \mathscr{F}, \mathbb{P})$, with a filtration $\{\mathscr{F}_t\}_{t\geq 0}$, see Definition 2.1, is given. As a function of x, u lies in a separable Hilbert space or a Banach space H and the initial data $u_0(\omega,x)$ is a $\mathscr{F}_0$ measurable random variable with values in H. The infinite-dimensional Brownian motion $B_t = (b_t^1, b_t^2, \ldots)$ (Wiener process) lies in another Hilbert space U and $C^{1/2} : U \to U_0$, a subset $U_0 \subset U$ defined by $U_0 = C^{1/2}U$, where C is the covariance matrix of B_t; see (1.57). Since B_t is infinite-dimensional, we will assume that C is trace class or that trace $C = \sum_{k\in\mathbb{Z}^3\setminus\{0\}} c_k < \infty$. The operator A is the generator of a strongly continuous semigroup e^{At}, $t \geq 0$, in H.

Definition 1.10. A stochastic process $u(\omega,x,t)$ is a mild solution of (1.67) if

$$u(t) = e^{At}u_0 + \int_0^t e^{A(t-s)}F(s,u(s))ds + \int_0^t e^{A(t-s)}G(s,u(s))dB_s, \tag{1.68}$$

$\mathbb{P}$-almost surely, and

$$\mathbb{P}\left(\int_0^t \|u\|^2(s)ds < \infty\right) = 1, \tag{1.69}$$

where $\|\cdot\|$ denotes Hilbert or Banach space norm.

We can then state the general existence theorem of mild solutions.

Theorem 1.9. *Suppose that F and G are deterministic maps from $[0,T] \times \Omega \times H$ into H and $L_2^0 = L_2(U_0,H)$, respectively. Furthermore assume that there exists a constant c such that*

1. $\|F(t,u) - F(t,w)\| + \|G(t,u) - G(t,w)\|_{L_2^0} \leq c\|u - w\|$
   and
2. $\|F(t,u)\|^2 + \|G(t,u)\|_{L_2^0}^2 \leq c^2(1 + \|u\|^2)$,

for all $t \in [0,T]$ and $\omega \in \Omega$. Then there exists a unique mild solution to (1.67); moreover, this solution has a continuous version among the equivalence class defined by (1.69):

1. For any $p \geq 2$, *there exists a constant* $C_{p,T}$ *such that*

$$\sup_{t\in[0,T]} E(\|u\|^p(t)) \leq C_{p,T}(1+E(\|u_0\|^p)). \tag{1.70}$$

2. For any $p > 2$, *there exists a constant* $\tilde{C}_{p,T}$ *such that*

$$E(\sup_{t\in[0,T]} \|u\|^p(t)) \leq \tilde{C}_{p,T}(1+E(\|u_0\|^p)). \tag{1.71}$$

It is of course not necessary to assume that F and G are deterministic; it suffices that they are measurable with respect to the correct σ-algebras; see [57].

Theorem 1.9 does not apply to (1.66). This is because the multiplicative noise term (1.66) involves jumps rather than Brownian motion. However, the theorem can be modified to give the local existence of solutions to (1.66). The existence of global (in t) solutions is much harder and can only be proven in special cases. One of these cases is explained in Chap. 4.

Chapter 2
Probability and the Statistical Theory of Turbulence

2.1 Ito Processes and Ito's Calculus

In this section we will define Ito processes that depend on the notion of the Ito integral. A very readable definition of the Ito integral and its properties can be found in Chap. 3 of Oksendal [51]. If $(\Omega, \mathscr{F}, \mathbb{P})$ is a probability space the following definition and example are also given in [51].

Definition 2.1. A filtration on $(\Omega, \mathscr{F})$ is a family $\{\mathscr{H}_t\}_{t\geq 0}$ of σ-algebras such that $\mathscr{H}_s \subset \mathscr{H}_t$, $0 \leq s \leq t$, or $\mathscr{H}_t$ is increasing. A n-dimensional stochastic process $\{M_t\}_{t\geq 0}$ on $(\Omega, \mathscr{F}, \mathbb{P})$ is called a martingale with respect to the filtration $\{\mathscr{H}_t\}_{t\geq 0}$ if:

1. M_t is $\mathscr{H}_t$-measurable for all t
2. $E(|M_t|) \leq \infty$ for all t
3. $E(M_t|\mathscr{H}_s) = M_s$ for all $t \geq s$

The expectation in (2) and the conditional expectations in (3) are the expectation with respect to the Gaussian density (1.51), and more information about the conditional expectation can be found in Appendix A in Oksendal [51].

Example 2.1. Brownian motion in $\mathbb{R}^n$ is a martingale with respect to the σ-algebras $\mathscr{F}_s$ generated by $\{b_s; s \leq t\}$:

1. Follows by definition.
2. $E(|b_t|)^2 = E(|b_t - b_0 + b_0|^2) = |b_0|^2 + nt$. Now use that the square root is an increasing function.
3. $E(b_t|\mathscr{F}_s) = E([b_t - b_s] + b_s|\mathscr{F}_s) = E(b_t - b_s|\mathscr{F}_s) + E(b_s|\mathscr{F}_s) = 0 + b_s$.

Now it turns out, see [51], that the Ito integral $\int f db_t$ can be defined as long as there exists an increasing filtration $\{\mathscr{H}_t\}_{t\geq 0}$ such that b_t is a martingale with respect to it and f is measurable with respect to $\mathscr{H}_t$. We then say that f is *adapted* to $\{\mathscr{H}_t\}_{t\geq 0}$.

We define an Ito process, following Oksendal [51].

B. Birnir, *The Kolmogorov-Obukhov Theory of Turbulence: A Mathematical Theory of Turbulence*, SpringerBriefs in Mathematics, DOI 10.1007/978-1-4614-6262-0_2,

Definition 2.2. Let b_t be a one-dimensional Brownian motion on the probability space $(\Omega, \mathscr{F}, \mathbb{P})$. An Ito process is a stochastic process on $(\Omega, \mathscr{F}, \mathbb{P})$ of the form

$$x_t = x_0 + \int_0^t u(x, \omega)\mathrm{d}s + \int_0^t w(s, \omega)\mathrm{d}b_s, \tag{2.1}$$

where there exists an increasing filtration $\{\mathscr{H}_t\}_{t \geq 0}$ such that b_t is a martingale with respect to it, w is $\mathscr{H}_t$ adapted, and

$$P\left(\int_0^t w(s, \omega)^2 \mathrm{d}s < \infty, \ \text{for all } t \geq 0\right) = 1.$$

We assume u is also $\mathscr{H}_t$ adapted and

$$P\left(\int_0^t |u(s, \omega)| \mathrm{d}s < \infty, \ \text{for all } t \geq 0\right) = 1.$$

We will use the shorthand

$$\mathrm{d}x_t = u(t, \omega)\mathrm{d}t + w(t, \omega)\mathrm{d}b_t \tag{2.2}$$

for an Ito process.

Now any Ito integral is a martingale with respect to the filtration $\{\mathscr{F}_t\}$ generated by $\{b_s; s \leq t\}$. Conversely, the martingale representation theorem, see [51], says that any martingale with respect to $\{\mathscr{F}_t\}$ can be written as an Ito integral.

The main computational tool in Ito's calculus is

Lemma 2.1 (One-Dimensional Ito's Formula). *Let x_t be the Ito process*

$$\mathrm{d}x_t = u\mathrm{d}t + w\mathrm{d}b_t,$$

and let $g(x,t) \in C^2([0,\infty) \times \mathbb{R})$ be twice continuously differentiable. Then

$$y_t = g(t, x_t)$$

is also an Ito process and

$$\mathrm{d}y_t = \frac{\partial g}{\partial t}(t, x_t)\mathrm{d}t + \frac{\partial g}{\partial x}(t, x_t)\mathrm{d}x_t + \frac{1}{2}\frac{\partial^2 g}{\partial x^2}(t, x_t)(\mathrm{d}x_t)^2, \tag{2.3}$$

where $(\mathrm{d}x_t)^2$ is computed by the rules

$$(\mathrm{d}t)^2 = \mathrm{d}t \cdot \mathrm{d}b_t = \mathrm{d}b_t \cdot \mathrm{d}t = 0, \ \text{but } (\mathrm{d}b_t)^2 = \mathrm{d}t.$$

We now give an example how one uses (2.3) to solve a differential equation.

Example 2.2 (Geometric Brownian Motion).
Let x_t be the Ito process $\mathrm{d}x_t = r\mathrm{d}t + \alpha\mathrm{d}b_t$ and solve the differential equation

$$\mathrm{d}z_t = z_t\mathrm{d}x_t = rz_t\mathrm{d}t + \alpha z_t\mathrm{d}b_t.$$

Dividing by z_t

$$\frac{\mathrm{d}z_t}{z_t} = r\mathrm{d}t + \alpha \mathrm{d}b_t = \mathrm{d}x_t$$

we see that a reasonable guess for the function g is

$$y_t = g(z_t) = \ln(z_t).$$

Applying Ito's formula (2.3), we get

$$\begin{aligned} \mathrm{d}y_t &= \frac{\partial g}{\partial x}(t, z_t)\mathrm{d}z_t + \frac{1}{2}\frac{\partial^2 g}{\partial x^2}(t, z_t)(\mathrm{d}z_t)^2 \\ &= \frac{z_t}{z_t}\mathrm{d}x_t - \frac{1}{2}\frac{(z_t)^2}{(z_t)^2}(\mathrm{d}x_t)^2 = \left(r - \frac{\alpha^2}{2}\right)\mathrm{d}t + \alpha \mathrm{d}b_t, \end{aligned}$$

since $\frac{\partial g}{\partial t} = 0$. Integrating

$$\mathrm{d}y_t = \mathrm{d}\ln(z_t) = \left(r - \frac{\alpha^2}{2}\right)\mathrm{d}t + \alpha \mathrm{d}b_t$$

with respect to t and exponentiating, we get that

$$z_t = z_0\, \mathrm{e}^{\{(r - \frac{\alpha^2}{2})t + \alpha b_t\}}.$$

This process is called geometric Brownian motion.

2.2 The Generator of an Ito Diffusion and Kolmogorov's Equation

In this section we will show how to use Ito processes to solve second-order partial differential equations following Oksendal [51].

Definition 2.3. A time-homogeneous Ito diffusion is a stochastic process $x_t : [0,\infty) \times \Omega \to \mathbb{R}^n$ satisfying the stochastic ordinary differential equation

$$\mathrm{d}x_t = u(x_t)\mathrm{d}t + w(x_t)\mathrm{d}b_t,\ x_0 = x, \tag{2.4}$$

where b_t is m-dimensional Brownian motion, $u : \mathbb{R}^n \to \mathbb{R}^n$, and $w : \mathbb{R}^n \to \mathbb{R}^{n\times m}$.

We can then define the generator of an Ito process.

Definition 2.4. Let x_t be a time-homogeneous Ito diffusion in $\mathbb{R}^n$. The infinitesimal generator A of x_t is defined by

$$Af(x) = \lim_{t\downarrow 0} \frac{E^x[f(x_t)] - f(x)}{t};\quad x \in \mathbb{R}^n. \tag{2.5}$$

The set of functions $f : \mathbb{R}^n \to R$ so that the limit exists at x is denoted $\mathscr{D}(A)(x)$, whereas $\mathscr{D}(A)$ is the domain of functions for which the limit exists for every $x \in \mathbb{R}^n$.

We now compute the generator of a time-homogeneous Ito diffusion.

Lemma 2.2. *Let x_t be the Ito diffusion*

$$\mathrm{d}x_t = u(x_t)\mathrm{d}t + w(x_t)\mathrm{d}b_t.$$

If $f \in C_0^2(\mathbb{R}^n)$ or f is twice differentiable with compact support, then $f \in \mathscr{D}_A$ and

$$Af(x) = \sum_{j=1}^{n} u_j(x)\frac{\partial f}{\partial x_j} + \frac{1}{2}\sum_{\{j,k\}=1}^{n} (ww^{\mathrm{T}})_{\{j,k\}}(x)\frac{\partial^2 f}{\partial x_j \partial x_k}. \tag{2.6}$$

We can then use Ito processes to solve partial differential equations.

Theorem 2.1 (Kolmogorov's Backward Equation). *Let $f \in C_0^2(\mathbb{R}^n)$ and consider*

$$u(x,t) = E^x(f(x_t)). \tag{2.7}$$

Then $u(\cdot,t) \in \mathscr{D}_A$ for each t and u is the unique solution of the initial value problem for the partial differential equation

$$\begin{aligned} \frac{\partial u}{\partial t} &= Au, \qquad t > 0,\ x \in \mathbb{R}^n, \\ u(x,0) &= f(x), \qquad x \in \mathbb{R}^n. \end{aligned} \tag{2.8}$$

Example 2.3 (The Heat Equation).
Consider the Ito process

$$\mathrm{d}x_t = \mathrm{d}b_t, \qquad \text{in } \mathbb{R}^n.$$

The generator of this process is

$$Af = \frac{1}{2}\Delta f$$

by Ito's formula. Thus by Theorem 2.1 the unique solution of the partial differential equation

$$\begin{aligned} \frac{\partial u}{\partial t} &= \frac{1}{2}\Delta u, \qquad t > 0, \qquad x \in \mathbb{R}^n \\ u(x,0) &= f(x), \qquad x \in \mathbb{R}^n \end{aligned}$$

is

$$u(x,t) = E^x(f(x_t)).$$

This is the probabilistic (Kac) solution of the heat equation.

2.2.1 The Feynman–Kac Formula

We now investigate how to solve a partial differential equation such as (2.8) that has an additional linear term (a potential).

Lemma 2.3 (The Feynman–Kac Formula). *Let $f \in C_0^2(\mathbb{R}^n)$ and assume that $q \in \mathbb{R}^n$ is bounded from below. Then the function*

$$u(x,t) = E^x(\mathrm{e}^{\{-\int_0^t q(x_s)\mathrm{d}s\}} f(x_t)) \tag{2.9}$$

satisfies $u(\cdot,t) \in \mathscr{D}_A$ for each t and u is the unique solution of the initial value problem for the partial differential equation

$$\frac{\partial u}{\partial t} = Au - qu, \qquad t > 0,\ x \in \mathbb{R}^n, \tag{2.10}$$

$$u(x,0) = f(x), \qquad x \in \mathbb{R}^n. \tag{2.11}$$

The upshot of the Feynman–Kac formula, Lemma 2.3, is that we can use the factor $\mathrm{e}^{\{-\int_0^t q(x_s)\mathrm{d}s\}}$ to remove the term $-qu$ from the equation and solve the usual (2.8) generated by the new Ito process $y_t = \mathrm{e}^{\{-\int_0^t q(x_s)\mathrm{d}s\}} x_t$.

2.2.2 Girsanov's Theorem and Cameron–Martin

We will show in this section how one can make a time change of a Brownian motion.

Theorem 2.2 (Girsanov's Theorem). *Let $y_t \in \mathbb{R}^n$ by an Ito process of the form*

$$\mathrm{d}y_t = u(t,\omega)\mathrm{d}t + \mathrm{d}b_t,\ t \le T,\ y_0 = 0,$$

where $T \le \infty$ is given and b_t is n-dimensional Brownian motion. Let

$$M_t = \mathrm{e}^{(\int_0^t u(s,\omega)\mathrm{d}b_s - \frac{1}{2}\int_0^t u^2(s,\omega)\mathrm{d}s)}, \tag{2.12}$$

where u satisfies Novikov's conditions

$$E(\mathrm{e}^{\{\frac{1}{2}\int_0^T u^2(s,\omega)\mathrm{d}s\}}) < \infty, \tag{2.13}$$

E being the expectation with respect to $\mathbb{P}$ the law of Brownian motion. Define the measure

$$\mathrm{d}Q(\omega) = M_T(\omega)\mathrm{d}\mathbb{P}(\omega); \tag{2.14}$$

then u_t is an n-dimensional Brownian motion with respect to the probability law Q for $t \le T$.

We can now use Cameron-Martin to remove the transport term $u \cdot \nabla w$ from the partial differential equation

$$\frac{\partial w}{\partial t} = -u \cdot \nabla w + \frac{1}{2}\Delta w.$$

Example 2.4 (Cameron–Martin). Let

$$y_t = M_t x_t = \mathrm{e}^{(-\int_0^t u(s,\omega)\mathrm{d}b_s - \frac{1}{2}\int_0^t |u(s,\omega)|^2 \mathrm{d}s)} x_t.$$

If x_t has the generator $A_1 = -u \cdot \nabla + \frac{1}{2}\Delta$, then y_t has the generator $A_2 = \frac{1}{2}\Delta$; now apply Theorem 2.1.

2.3 Jumps and Lévy Processes

We will now define stochastic processes with jumps, following Oksendal and Sulem [52], where more information can be found. A Lévy process is a stochastic process on a filtered probability space $(\Omega, \mathscr{F}, \{\mathscr{F}_t\}_{t\geq 0}, \mathbb{P})$ that takes its values in $\mathbb{R}$ and is continuous in probability and has stationary independent increments. One can always find a version of η_t that is right continuous with limits from the left; such processes are called *cadlag*. η_t is $\{\mathscr{F}_t\}$ adapted and $\eta_0 = 0$ almost surely.

Definition 2.5. Let b_t be a one-dimensional Brownian motion on the probability space $(\Omega, \mathscr{F}, \mathbb{P})$ and suppose that $N(t,z)$ is the number process of a Lévy process η_t. An Ito–Lévy process is a stochastic process on $(\Omega, \mathscr{F}, \mathbb{P})$ of the form

$$\mathrm{d}x_t = u(t,\omega)\mathrm{d}t + w(t,\omega)\mathrm{d}b_t + \int_{\mathbb{R}} \gamma(t,z,\omega)\bar{N}(\mathrm{d}z,\mathrm{d}t). \tag{2.15}$$

$\bar{N}$ is called the compensated jump measure of η_t,

$$\bar{N}(\mathrm{d}z,\mathrm{d}t) = N(\mathrm{d}z,\mathrm{d}t) - m(\mathrm{d}z)\mathrm{d}t, \text{ if } |z| < R,$$

and

$$\bar{N}(\mathrm{d}z,\mathrm{d}t) = N(\mathrm{d}z,\mathrm{d}t), \text{ if } |z| \geq R,$$

where $m(U) = \int_U E(N(\mathrm{d}z,1))$ is the so-called Lévy measure of η_t; see Sect. 1.6.

The main computational tool in Ito's calculus for Ito–Lévy processes is

Lemma 2.4 (One-Dimensional Ito's Formula). *Let x_t be the Ito–Lévy process*

$$\mathrm{d}x_t = u\mathrm{d}t + w\mathrm{d}b_t + \int_{\mathbb{R}} h(z,t)\bar{N}(\mathrm{d}z,\mathrm{d}t).$$

Let $g(x,t) \in C^2([0,\infty) \times \mathbb{R})$ be twice continuously differentiable. Then

$$y_t = g(t,x_t)$$

is also an Ito–Lévy process and

$$
\begin{aligned}
\mathrm{d}y_t = {} & \frac{\partial g}{\partial t}(t,x_t)\mathrm{d}t + \frac{\partial g}{\partial x}(t,x_t)(u\mathrm{d}t + w\mathrm{d}b_t) + \frac{1}{2}\frac{\partial^2 g}{\partial x^2}(t,x_t)(u\mathrm{d}t + w\mathrm{d}b_t)^2 \\
& + \int_{z<R}\left(g(t,x_t+\gamma(t,z)) - g(t,x_t) - \frac{\partial g}{\partial x}(t,x_t)\gamma(t,x)\right) m(\mathrm{d}z)\mathrm{d}t \\
& + \int_{\mathbb{R}} (g(t,x_t+\gamma(t,z)) - g(t,x_t))\,\bar{N}(\mathrm{d}z,\mathrm{d}t), \qquad (2.16)
\end{aligned}
$$

where $(u\mathrm{d}t + w\mathrm{d}b_t)^2$ *is computed by the rules*

$$(\mathrm{d}t)^2 = \mathrm{d}t\cdot\mathrm{d}b_t = \mathrm{d}b_t\cdot\mathrm{d}t = 0, \text{ but } (\mathrm{d}b_t)^2 = \mathrm{d}t.$$

We now give an example how one uses (2.16) to solve a differential equation.

Example 2.5 (Geometric Lévy Process).
We solve the differential equation

$$\mathrm{d}z_t = z_t[r\mathrm{d}t + \alpha\mathrm{d}b_t + \int_{\mathbb{R}} h(z,t)\bar{N}(\mathrm{d}z,\mathrm{d}t)].$$

Dividing by z_t

$$\frac{\mathrm{d}z_t}{z_t} = r\mathrm{d}t + \alpha\mathrm{d}b_t + \int_{\mathbb{R}} h(z,t)\bar{N}(\mathrm{d}z,\mathrm{d}t),$$

we see that a reasonable guess for the function g is

$$y_t = \ln(z_t).$$

Applying Ito's formula (2.16), we get

$$
\begin{aligned}
\mathrm{d}y_t &= \frac{\partial \ln(z_t)}{\partial x}(t,x_t)z_t(r\mathrm{d}t + \alpha\mathrm{d}b_t) + \frac{1}{2}\frac{\partial^2 \ln(z_t)}{\partial x^2}(t,x_t)(z_t)^2(r\mathrm{d}t + \alpha\mathrm{d}b_t)^2 \\
&\quad + \int_{\mathbb{R}} \ln(1+h(t,z))\bar{N}(\mathrm{d}z,\mathrm{d}t) + \int_{\mathbb{R}} (\ln(1+h(t,z)) - h(t,z))\,m(\mathrm{d}z)\mathrm{d}t \\
&= \frac{z_t}{z_t}(r\mathrm{d}t + \alpha\mathrm{d}b_t) - \frac{1}{2}\frac{(z_t)^2}{(z_t)^2}(r\mathrm{d}t + \alpha\mathrm{d}b_t)^2 + \int_{\mathbb{R}} \ln(1+h(t,z))\bar{N}(\mathrm{d}z,\mathrm{d}t) \\
&\quad + \int_{\mathbb{R}} (\ln(1+h(t,z)) - h(t,z))\,m(\mathrm{d}z)\mathrm{d}t \\
&= \left(r - \frac{\alpha^2}{2}\right)\mathrm{d}t + \alpha\mathrm{d}b_t + \int_{\mathbb{R}} \ln(1+h(t,z))\bar{N}(\mathrm{d}z,\mathrm{d}t) \\
&\quad + \int_{\mathbb{R}} (\ln(1+h(t,z)) - h(t,z))\,m(\mathrm{d}z)\mathrm{d}t,
\end{aligned}
$$

since $\frac{\partial g}{\partial t} = \frac{\partial \ln(z_t)}{\partial t} = 0$. Integrating

$$dy_t = d\ln(z_t) = \left(r - \frac{\alpha^2}{2}\right) dt + \alpha db_t + \int_{\mathbb{R}} \ln(1+h(s,z))\bar{N}(dz,ds)$$
$$+ \int_{\mathbb{R}} (\ln(1+h(s,z)) - h(s,z))\, m(dz)$$

with respect to t and exponentiating, we get that

$$z_t = z_0\, e^{\{(r-\frac{\alpha^2}{2})t+\alpha b_t+\int_0^t\int_{\mathbb{R}}\ln(1+h(s,z))\bar{N}(dz,ds)+\int_0^t\int_{\mathbb{R}}(\ln(1+h(s,z))-h(s,z))m(dz)ds\}}.$$

This process is called the geometric Lévy process.

2.4 Spectral Theory for the Operator K

We write the stochastic Navier–Stokes equation in integral form

$$u = e^{K(t)} e^{\int_0^t dq} M_t u^0 + \sum_{k\neq 0} c_k^{1/2} \int_0^t e^{K(t-s)} e^{\int_s^t dq} M_{t-s} db_s^k e_k(x)$$
$$+ \sum_{k\neq 0} d_k \int_0^t e^{K(t-s)} e^{\int_s^t dq} M_{t-s} |k|^{1/3} dt\; e_k(x), \tag{2.17}$$

where K is the linear (Navier–Stokes) operator

$$K = \nu\Delta + \nabla\Delta^{-1}\mathrm{tr}(\nabla u \nabla),$$
$$M_t = \exp\{-\int u(B_s,s)dB_s - \frac{1}{2}\int_0^t |u(B_s,s)|^2 ds\}$$

is a martingale with $B_t \in \mathbb{R}^3$ an auxiliary Brownian motion, and

$$3\int_s^t dq = \sum_{k\neq 0}^{m} \left\{ \int_0^t \int_{\mathbb{R}} \ln(1+h_k)\bar{N}^k(ds,dz) + \int_0^t \int_{\mathbb{R}} (\ln(1+h_k) - h_k) m_k(ds,dz) \right\},$$

by Ito's formula and a computation similar to the one that produces the geometric Lévy process; see [52]. We have set the rates $\eta_k = |k|^{1/3}$ since the two terms in the additive noise are really two parts of the same noise term; see Sect. 1.7. The operator K does not generate a semi-group because of its dependence on u, but with some conditions on u, see below, it generates a flow. The notation $e^{K(t-s)} f(s)$ simply means that we solve the equation $f_t = Kf$, with initial data $f(s)$ for the time interval $[s,t]$; see Lemma 2.6 below.

The form of the integral equation (2.17) requires a couple of assumptions. The first observation is that the pressure term $\nabla\Delta^{-1}\mathrm{tr}(\nabla u \cdot \nabla \cdot)$ is independent of the fluid velocity $u(x,t)$ at the point x. This is of course true since x is a set of measure zero and we can be set the integrand to any value at x without changing the integral. In other words, the pressure gradient can be treated as a global force that depends on the velocity field as a whole not only on some particular fluid particle. This is consistent with the view of pressure in most of fluid dynamics. The other assumption is that pressure acts as additional diffusion and the integral equation (2.17) describes a (Ito) diffusion. This is also consistent with most researchers view of pressure but seems to be a more radical assumption from a mathematical point of view. However, it can be proven to be true using the vorticity formulation of the Navier–Stokes equation; see Sect. 3.7. The first assumption implies that the right-hand side of (2.17) is independent of $u(x,t)$ so that by Ito's formula the integral equation (2.17) is equivalent to the initial value problem (1.65). The second assumption implies that we can apply Girsanov's Theorem 2.2 to remove the inertial (drift) term from the linearized Navier-Stokes operator in lieu of the martingale M_t.

To proceed we need to develop the spectral theory of the operator K. The existence of unique turbulent solutions to the stochastic Navier–Stokes equation (1.65) can be proven in some cases. For example, if the equation is driven by a strong swirling flow, see [17] and Chap. 4. This result is not terribly surprising. If the initial data had the symmetry of the swirl then the deterministic problem would be two-dimensional and the global existence of the two-dimensional Navier–Stokes equation is well known. It is also well known that if the initial data is close to such a two-dimensional flow then global existence can be extended to this case also; see [3, 4] for another such example.

In [17] the author obtained the global bound for the Sobolev space norm of u, based on $L^2(\mathbb{T}^3)$ with index $\frac{11}{6}^+ = \frac{11}{6} + \varepsilon$, ε small, for a swirling flow:

$$E(\|u\|^2_{\frac{11}{6}^+}(t)) \leq C, \tag{2.18}$$

where E denotes the expectation and the constant C is independent of t. The Sobolev space consists of Hölder continuous functions of Hölder index $1/3$, as pointed out by Onsager [54]. Sobolev spaces are defined in Sect. 4.1.

We will now derive a bound for the pressure operator. Consider the linearized Navier–Stokes equation that will be discussed in more detail in Sect. 2.8:

$$z_t = B(u)z = \nu\Delta z - u\cdot\nabla z - z\cdot\nabla u + 2\nabla\Delta^{-1}\mathrm{tr}(\nabla u\cdot\nabla z) = \bar{K}(u)z - u\cdot\nabla z - z\cdot\nabla u. \tag{2.19}$$

We start with a standard estimate of the pressure term and let $\frac{p}{q}^+$ denote a real number strictly greater than $\frac{p}{q}$.

Lemma 2.5. *Let $Dw = 2\nabla\Delta^{-1}\mathrm{tr}\nabla u\cdot\nabla w$. Then*

$$|Dw|_2 \leq C\|u\|_{\frac{3}{2}^+}|w|_2 \tag{2.20}$$

and for functions with mean zero

$$|Dw|_2 \leq C|\nabla u|_2|\nabla w|_2, \tag{2.21}$$

where C is a constant.

Proof.

$$\begin{aligned}\frac{1}{2}|Dw|_2 \leq |\nabla\Delta^{-1}\mathrm{tr}\nabla u\cdot\nabla w|_2 \leq \|\nabla u\cdot\nabla w\|_{(-1,2)} \leq \|\nabla u\|_{(-1,4)}\|\nabla w\|_{(-1,4)}\\ \leq C\|\nabla u\|_{(-\frac{1}{4}^+,2)}\|\nabla w\|_{(-\frac{1}{4}^+,2)} \leq C|\nabla u|_2|\nabla w|_2\end{aligned}$$

by Schwartze's and Sobolev's inequalities and the Sobolev imbedding. The last inequality only holds for functions with means zero. Moreover,

$$\begin{aligned}\frac{1}{2}|Dw|_2 \leq |\nabla\Delta^{-1}\mathrm{tr}\nabla u\cdot\nabla w|_2 \leq \|\nabla u\cdot\nabla w\|_{(-1,2)} \leq \|\nabla u\|_{(-1+\alpha,4)}\|\nabla w\|_{(-1-\alpha,4)}\\ \leq C\|\nabla u\|_{(\alpha-\frac{1}{4}^+,2)}\|\nabla w\|_{(-\alpha-\frac{1}{4}^+,2)} \leq C\|u\|_{\frac{3}{2}^+}|w|_2,\end{aligned}$$

by the same inequalities as above, the duality of the Sobolev spaces with a positive and a negative index and by choosing $\alpha = \frac{3}{4}^+$.

Lemma 2.6. *Suppose that*

$$E(\|u\|^2_{\frac{3}{2}^+}) \leq C, \tag{2.22}$$

then the operator B, in (2.19), generates a flow for initial data $f_0 \in L^2(\mathbb{T}^3)$ denoted by e^{Bt}. The same conclusion holds for the operator $\bar{K}$ in (2.19).

Proof. The proof follows the proof of Lemma 7.2 in [16]. Consider the operators $Aw = \nu\Delta w$ and

$$Sw = -u\cdot\nabla w - w\cdot\nabla u + 2\nabla\Delta^{-1}\mathrm{tr}(\nabla u\cdot\nabla w).$$

A generates a contraction semigroup and we now show that S is A bounded or that there exists a constant C such that

$$|Sw|_2 \leq C|w|_2 + \frac{1}{2}|Aw|_2.$$

By Minkowski's inequality

$$\begin{aligned}|Sw|_2 &\leq |u\cdot\nabla w|_2 + |w\cdot\nabla u|_2 + 2|\nabla\Delta^{-1}\mathrm{tr}(\nabla u\cdot\nabla w)|_2\\ &\leq |u|_\infty|\nabla w|_2 + |\nabla u|_2|w|_\infty + C\|u\|_{\frac{3}{2}^+}|w|_2\\ &\leq C\|u\|_{\frac{3}{2}^+}\|w\|_1 + C|\nabla u|_2\|w\|_{\frac{3}{2}^+}\end{aligned}$$

by Lemma 2.5 and Sobolev's inequalities

$$\leq C\|u\|_{\frac{3}{2}^+}|w|_2 + \frac{1}{2}|\nu\Delta w|_2$$

by interpolation. Moreover in the space of divergence-free function S is dissipative, namely,

$$\langle w, Sw\rangle = -\langle w, u\cdot\nabla w\rangle - \langle w, w\cdot\nabla u\rangle + 2\langle w, \nabla\Delta^{-1}\mathrm{tr}(\nabla u\nabla w)\rangle = 0$$

by use of the periodic boundary conditions. The remainder of the proof follows the proof of Lemma 7.2 in [16]. The last inequality implies that

$$\|Sw\|_2 \le C'\|w\|_2 + \frac{1}{2}\|\nu\Delta w\|_2,$$

where the norm is defined by $\|w\| = \sqrt{E(|w|_2^2)}$, and we have used the hypothesis (2.18) and that A is deterministic. This shows that S is A bounded in the (probability) space $L^2((\Omega,\mathscr{F},\mathbb{P});L^2(\mathbb{T}^3))$; see Sect. 4.1 for a definition of this space. Since S is dissipative and A bounded $B = A+S$ also generates a flow e^{Bt} in this space, see [32]. The same conclusion follows for the operator $\bar{K}$ in (2.19) (and K) by setting $S = D$, where D is the pressure operator in Lemma 2.5.

Then using the bound (2.18), we get a (spectral) estimate on the operator K.

Lemma 2.7. *Suppose that (2.18) holds, then the pressure operator is bounded by the spectrum of a symmetric operator with discrete spectrum λ_k^2 and satisfies the estimate*

$$-C|k|^{2/3} \le -\lambda_k \le |\nabla\Delta^{-1}\mathrm{tr}\nabla u \cdot \nabla P_k|_2 \le \lambda_k \le C|k|^{2/3},\ k \in \mathbb{Z}^3, \tag{2.23}$$

on the Hilbert space $H^{\frac{11}{6}^+}(\mathbb{T}^3)$, in the inertial range; see below. P_k is the projection onto the kth eigenspace of the symmetric operator. Moreover, in the inertial range, the operator K satisfies the bound

$$-C|k|^{2/3} - 4\nu\pi^2|k|^2 \le |KP_k|_2 \le C|k|^{2/3} + 4\nu\pi^2|k|^2,\ k \in \mathbb{Z}^3. \tag{2.24}$$

We will use this estimate below in order to compute the structure functions of turbulence or the moments of the velocity difference at two points in the fluid, in the inertial range of turbulence, where $1/L \le |k| \le 1/\eta$, $k_0 = 1/\eta = (\varepsilon/\nu^3)^{1/4}$, a constant. $\eta = 1/k_0$ is called the Kolmogorov length scale, ε is the energy dissipation rate (1.8), and L is a typical length scale associated with the large eddies in the flow. The above estimate implies that for a large Reynolds number where ν is small and $1/L \le |k| \le 1/\eta$, we can think of the spectrum of K growing as a constant times $|k|^{2/3}$, with the error $4\nu\pi^2|k|^2$, in the inertial range.

The proof of Lemma 2.7 and the bounds (2.23) and (2.24) is the following.

Proof. A general vector w in $L^2(\mathbb{T}^3)$ can be decomposed into a divergence-free and an irrotational part

$$w = u + v = \nabla\times A + \nabla\phi,$$

respectively. The pressure operator $Df = \nabla\Delta^{-1}\mathrm{tr}\nabla u\cdot\nabla f$ maps the subspace U of divergence-free vectors in $L^2(\mathbb{T}^3)$ to the subspace of the irrotational vectors V in $L^2(\mathbb{T}^3)$. Thus D has no eigenvalues or eigenvectors in U. However, the magnitude of the pressure gradient, the force that keeps the fluid velocity in U, is measured by the norm $|Df|_2$ or by

$$|Df|_2^2 = \langle Df, Df\rangle = \langle f, D^{\mathrm{T}}Df\rangle,$$

where D^{T} is the transpose of D on V. Thus the magnitude of D is measured by $|\lambda_k|$ where the λ_k^2 are the eigenvalues of the symmetric operator $D^{\mathrm{T}}D$ on the eigenspaces P_k in U, if $D^{\mathrm{T}}D$ has discrete spectrum. We will establish the discreteness of the spectrum and estimate the spectrum of $D^{\mathrm{T}}D$ by comparing it with the spectrum of the symmetric operator $(\partial_x^{2/3})^2$ on U. For $f \in H^{2/3}$, the Sobolev space based on $L^2(\mathbb{T}^3)$ with index $2/3$, see Sect. 4.1, D satisfies the estimate

$$|Df|_2 \leq C\|u\|_{\frac{11}{6}^+}|\partial_x^{2/3}f|_2. \tag{2.25}$$

The estimate (2.25) follows from Fourier transform

$$\begin{aligned}
\widehat{Df} &= \nabla\Delta^{-1}\widehat{\mathrm{tr}\nabla u}\cdot\nabla f = \frac{2\pi i k}{|k|^2}\mathrm{tr}\sum_{j\neq 0}(k-j)\otimes\hat{u}(k-j)j\otimes\hat{f}(j) \\
&\leq 2\pi\frac{1}{|k|^{3/2}}\mathrm{tr}\sum_{j\neq 0}|k|^{1/2}|j|^{1/3}|k-j||\hat{u}(k-j)||j|^{2/3}|\hat{f}(j)| \\
&\leq \frac{1}{(2\pi)^{3/2}|k|^{3/2^+}}\left(\sum_{j\neq 0}|\widehat{\partial_x^{\frac{11}{6}^+}u}(k-j)|^2\right)^{1/2}\left(\sum_{j\neq 0}|\widehat{\partial_x^{2/3}f}(j)|^2\right)^{1/2}
\end{aligned}$$

by Schwartz's inequality. Now squaring and summing in k we get (2.25).

Thus for nondegenerate fluid velocities u that satisfy (2.18), $D^{\mathrm{T}}D$ maps a dense subset of $H^{2/3}(\mathbb{T}^3)\cap U$ onto $L^2(T^3)\cap U$. This means that the resolvent $(I-D^{\mathrm{T}}D)^{-1}$ maps $L^2(\mathbb{T}^3)\cap U$ onto $H^{2/3}(\mathbb{T}^3)\cap U$. Since the latter space sits compactly in the former, $(I-D^{\mathrm{T}}D)^{-1}$ is a compact operator with discrete spectrum. This implies that $D^{\mathrm{T}}D$ also has discrete spectrum.

The estimate (2.23) follows from the minimax principle, see [32], comparing the eigenvalues of the symmetric operators

$$D^{\mathrm{T}}D \leq C^2\|u\|^2_{\frac{11}{6}^+}(\partial_x^{2/3})^2$$

and taking both branches of the square root. Similarly, (2.23) follows by comparing the eigenvalues of the symmetric operators:

$$(-\nu\Delta+D)^{\mathrm{T}}(-\nu\Delta+D) = \nu^2\Delta^2-\nu(D^{\mathrm{T}}\Delta+\Delta D)+D^{\mathrm{T}}D \leq (C\|u\|_{\frac{11}{6}^-}\partial_x^{2/3}-\nu\Delta)^2.$$

This concludes the proof of Lemma 2.7.

2.5 The Feynman–Kac Formula and the Log-Poissonian Processes

The processes found by She and Leveque [64], and shown to be log-Poisson processes by She and Waymire [65] and by Dubrulle [23], are produced by applying the

Feynman–Kac formula, Lemma 2.3, to the potential dq. Namely, $e^{\int_0^t dq} = e^{\sum_{k\neq 0}^m \int_0^t dq_k}$ and by setting $h_k = \beta - 1$ and computing the mean of N_t^k

$$E(N_t^k) = \int_{\mathbb{R}} m_k(t, dz) = -\frac{\gamma \ln|k|}{\beta - 1}, \tag{2.26}$$

we get that

$$\begin{aligned} 3\int_0^t dq_k &= \int_0^t \int_{\mathbb{R}} \ln(1+h_k)\bar{N}^k(ds, dz) + \int_0^t \int_{\mathbb{R}} (\ln(1+h_k) - h_k) m_k(ds, dz) \\ &= N_k(t)\ln(\beta) + (\beta - 1)\left(\gamma \frac{\ln|k|}{\beta - 1}\right). \end{aligned}$$

The upshot is the term

$$e^{\int_0^t dq_k} = e^{(\gamma \ln|k| + N_k \ln\beta)/3} = \left(|k|^\gamma \beta^{N_k}\right)^{1/3} = \left(|k|^\gamma \beta^{N_t^k}\right)^{1/3} \tag{2.27}$$

in the (implicit) solution (2.17) of the stochastic Navier–Stokes equation. These are exactly the log-Poisson processes found by the above authors. This gives

$$\ln E((e^{\gamma \ln|k| + N_k \ln\beta})^{\frac{p}{3}}) = \ln E((|k|^\gamma \beta^{N_k})^{\frac{p}{3}}) = \gamma\left(\frac{p}{3} - \frac{\beta^{p/3} - 1}{\beta - 1}\right) \ln|k| = -\tau_p \ln|k|, \tag{2.28}$$

for the logarithm of the pth moment, where τ_p are the intermittency corrections in (2.35). Now the expression

$$\tau_p = -\gamma\left(\frac{p}{3} - \frac{\beta^{p/3} - 1}{\beta - 1}\right)$$

implies that $\tau_0 = 0$ and $\tau_3 = 0$ independently of γ. The latter condition is required by the Kolmogorov 4/5th law; see [26]. However, to be consistent with the spectral theory of the operator D above, that moves energy around in quanta of $|k|^{2/3}$, we should set $\gamma = 2/3$. This means that the log-Poissionian processes also move energy in quanta of $|k|^{2/3}$ in Fourier space. However, $|k|^{2/3}$ is multiplied by $\beta^{N_t^k}$ in (2.27) above, namely, the number of jumps on the kth level contributes to the transfer of energy, and so far β is a free parameter. We follow [64] in making the assumption that determines β; see also [66]. The basic assumption is that most singular structures in the turbulent fluid are one-dimensional vortex lines (filaments) that the highest moments capture. Thus (assuming $0 < \beta < 1$) by the Lagrange transformation, see [64],

$$\tau_p = -\frac{2}{3}\left(\frac{p}{3}\right) + \frac{2}{3}\frac{1}{1-\beta} - \frac{2}{3}\frac{\beta^{p/3}}{1-\beta} \to -\frac{2}{3}\left(\frac{p}{3}\right) + \frac{2}{3}\frac{1}{1-\beta} = -\frac{2}{3}\left(\frac{p}{3}\right) + C_0$$

as $p \to \infty$, where $C_0 = 2$ is the codimension of the one-dimensional vortex lines and this implies that $\beta = 2/3$. We will make this choice of β.

Thus we see that the jumps multiplying u in (1.65) produce the log-Poisson processes $(|k|^{\frac{2}{3}}\left(\frac{2}{3}\right)^{N_t^k})^{\frac{1}{3}}$ in the integral equation for u:

$$u = \mathrm{e}^{K(t)} \left(\prod_k^m |k|^{\frac{2}{3}} (2/3)^{N_t^k} \right)^{\frac{1}{3}} M_t u_0$$
$$+ \sum_{k \neq 0} c_k^{1/2} \int_0^t \mathrm{e}^{K(t-s)} \left(\prod_j^m |j|^{\frac{2}{3}} (2/3)^{N_{(t-s)}^j} \right)^{\frac{1}{3}} M_{t-s} \mathrm{d}b_s^k e_k(x)$$
$$+ \sum_{k \neq 0} \mathrm{d}_k \int_0^t \mathrm{e}^{K(t-s)} \left(\prod_j^m |j|^{\frac{2}{3}} (2/3)^{N_{(t-s)}^j} \right)^{\frac{1}{3}} M_{t-s} |k|^{1/3} \mathrm{d}t \; e_k(x)$$

since only the kth log-Poissonian processes are correlated with the kth Fourier component. This equation clearly shows how the intermittency in the velocity (in (1.65)) causes intermittency in the dissipation through the Navier–Stokes evolution, if we recall how the discrete (Poisson) distribution picks the kth term (associated with e_k) out of the product. M_t is the martingale

$$M_t = \exp\left\{ - \int_0^t u(B_s, s) \cdot \mathrm{d}B_s - \frac{1}{2} \int_0^t |u(B_s, s)|^2 \mathrm{d}s \right\}. \tag{2.29}$$

It is the Radon–Nikodym derivative of the measure, see Theorem 2.2, of the associated Ito process

$$\mathrm{d}x_t = -u\mathrm{d}t + \sqrt{2\nu}\mathrm{d}B_t,$$

u being the fluid velocity and $B_t = (B_t^1, B_t^2, B_t^3)$ a three-dimensional vector of auxiliary Brownian motions, with respect to the usual Brownian measure.

2.6 The Kolmogorov–Obukhov–She–Leveque Theory

In 1941 Kolmogorov and Obukhov [34, 35, 49] proposed a statistical theory of turbulence based on dimensional arguments. The main consequence and test of this theory was that the structure functions of the velocity differences of a turbulent fluid

$$E(|u(x,t) - u(x+l,t)|^p) = S_p = C_p l^{p/3}$$

should scale with the distance (lag variable) l between them, to the power $p/3$. This theory was immediately criticized by Landau for not taking into account the influence of the large flow structure on the constants C_p and later for not including the influence of the intermittency in the velocity fluctuations on the scaling exponents; see [2].

In 1962 Kolmogorov and Obukhov [36, 50] proposed a corrected theory were both of the above issues were addressed. They presented their refined similarity hypothesis

$$S_p = C'_p < \tilde{\varepsilon}^{p/3} > l^{p/3}, \tag{2.30}$$

where l is the lag variable and the averaged energy dissipation rate is

$$\tilde{\varepsilon} = \frac{1}{\frac{4}{3}\pi l^3} \int_{|s|\leq l} \varepsilon(x+s)\mathrm{d}s, \tag{2.31}$$

ε being the mean energy dissipation rate (1.8). They also pointed out that the scaling exponents for the first two structure functions could be corrected by log-normal processes. However, for higher-order structure functions, the log-normal processes gave intermittency corrections inconsistent with contemporary simulations and experiments.

In the refined similarity hypothesis (2.30) the averaged dissipation rate $\tilde{\varepsilon}$ will depend on the large flow structure, so its addition addresses Landau's objections at least partially. The assumption is that

$$< \tilde{\varepsilon}^{p/3} > \sim l^{\tau_p},$$

because of intermittency, where the τ_p are called the intermittency corrections (to the scaling). Consequently, intermittency corrections are also produced:

$$S_p = C'_p < \tilde{\varepsilon}^{p/3} > l^{p/3} = C_p l^{p/3+\tau_p} = C_p l^{\zeta_p},$$

where

$$\zeta_p = \frac{p}{3} + \tau_p$$

are the scaling exponents, with intermittency corrections included, and the C_ps are not universal but depend on the large flow structure. We will see below that starting with (1.65), this scaling hypothesis in fact holds.

The She–Leveque intermittency corrections are

$$\tau_p = -\frac{2p}{9} + 2(1-(2/3)^{p/3}),$$

given by the log-Poissonian processes derived above. These intermittency corrections are consistent with contemporary simulations and experiments; see [2, 59, 64, 66].

2.7 Estimates of the Structure Functions

We will now show how the integral form (2.17) of the stochastic Navier-Stokes equation can be used to compute an estimate for the structure functions of turbulence.

In order to compute the structure functions of turbulence or the moments of the velocity difference at two points in the fluid, we need to estimate the operator K above and compare (2.23). Recall the eigenvalues $\lambda_k > 0$ that are the square roots of the eigenvalues of the symmetric operator $D^T D$ above, with P_k the projector onto the corresponding eigenspace. Then (2.24) can be reformulated as

$$\begin{aligned} -C|k|^{2/3} - 4\nu\pi^2|k|^2 &\le -\lambda_k - \nu 4\pi^2|k|^2 \le |KP_k|_2 \\ &\le \lambda_k + \nu 4\pi^2|k|^2 \le C|k|^{2/3} + \nu 4\pi^2|k|^2, \end{aligned} \tag{2.32}$$

if u satisfies the bound

$$E(\|u\|_{\frac{11}{6}^+})(t) \le C, \tag{2.33}$$

by the above. For a large Reynolds number ν is small and since $|k|^2 \le k_0^2$, $k_0 = (\varepsilon/\nu^3)^{1/4}$, where k_0 is the inverse of the Kolmogorov length, we can now think of the spectrum of K growing as a constant times $|k|^{2/3}$ in the inertial range. ε is the dissipation rate (1.8). The coefficient C is a constant times a Sobolov space norm of u, by the estimate (2.25); see [17].

Now estimates of the structure function are possible and we get the following result. Suppose that the coefficients c_k and d_k in (1.65) satisfy the conditions $\sum_{k\in\mathbb{Z}^3\setminus\{0\}} c_k < \infty$ and $\sum_{k\in\mathbb{Z}^3\setminus\{0\}} |k|^{1/3}|\mathrm{d}_k| < \infty$. Then the scaling of the structure functions of (1.65) is

$$S_p \sim C_p|x-y|^{\zeta_p}, \tag{2.34}$$

where

$$\zeta_p = \frac{p}{3} + \tau_p = \frac{p}{9} + 2(1-(2/3)^{p/3}), \tag{2.35}$$

$\frac{p}{3}$ being the Kolmogorov–Obukhov '41 scaling and τ_p the She–Leveque intermittency corrections, when the lag variable $|x-y|$ is small.

The values in (2.35) agree with experimental values in [59]; they are in agreement with the Kolmogorov–Obukhov scaling hypothesis with intermittency corrections, computed by She and Leveque, but disagree with the log-normal distribution [36, 50], for the intermittency corrections.

The estimate of the first structure function is straightforward:

$$\begin{aligned} S_1(x,y,t) &= E(|u(x,t)-u(y,t)|) \\ &= 2\sum_{k\in\mathbb{Z}^3\setminus\{0\}} \mathrm{d}_k \int_0^t \mathrm{e}^{-\lambda_k(t-s)}|k|^{1/3}\mathrm{d}s\, E([\mathrm{e}^{\gamma\ln|k|+N_k\ln(\beta)}]^{1/3})\sin(\pi k\cdot(x-y)) \\ &\le \frac{2}{C}\sum_{k\in\mathbb{Z}^3\setminus\{0\}} |\mathrm{d}_k|\frac{(1-\mathrm{e}^{-\lambda_k t})}{|k|^{\zeta_1}}|\sin(\pi k\cdot(x-y))|. \end{aligned} \tag{2.36}$$

We have estimated $K(t)$ by $\lambda_k = C|k|^{2/3}$ in the second line (we use this approximation, $\nu = 0$, throughout the computations) and also used the expectation of the Poisson jump process:

$$E([\mathrm{e}^{\gamma \ln|k| + N_k \ln(\beta)}]^{1/3}) = \frac{1}{|k|^{\tau_1}}.$$

We used the lower estimate in (2.32) and this makes the estimate in (2.36) be an overestimate of the efficiency of the cascade. The measure of the discrete process must be written as

$$\sum_{l=-\infty}^{\infty} \delta_{l,k} \prod_{j \neq l}^{m} \delta_{N_t^j} \sum_{j=0}^{\infty} (\cdot) \frac{m_l^j}{j!} \mathrm{e}^{(-m_l)}, \tag{2.37}$$

where $\delta_{l,k} = 0, l \neq k, 1, l = k$ is the Kronecker delta function, because N_t^k depends on the kth Fourier component e_k (or $\mathrm{d}b_t^k$ and $|k|^{1/3}\mathrm{d}t$) but is independent of the components with different wave numbers. The δ functions in the product imply that the probabilities of all the N_t^js, $j \neq k$ concentrate at 0.

Now, if $\sum_{k \in \mathbb{Z}^3 \setminus \{0\}} |\mathrm{d}_k| < \infty$, then we get a stationary state as $t \to \infty$:

$$S_1(x,y,\infty) \leq \frac{2}{C} \sum_{k \in \mathbb{Z}^3 \setminus \{0\}} \frac{|\mathrm{d}_k|}{|k|^{\zeta_1}} |\sin(\pi k \cdot (x-y))|,$$

and for $|x-y|$ small

$$S_1(x,y,\infty) \sim \frac{2\pi^{\zeta_1}}{C} \sum_{k \in \mathbb{Z}^3 \setminus \{0\}} |\mathrm{d}_k||x-y|^{\zeta_1},$$

where $\zeta_1 = 1/3 + \tau_1 \approx 0.37$.

A similar computation gives the second structure function:

$$\begin{aligned} S_2 &= E(|u(x,t) - u(y,t)|^2) \\ &\leq \frac{2}{C} \sum_{k \in \mathbb{Z}^3 \setminus \{0\}} c_k \frac{1 - \mathrm{e}^{-2\lambda_k t}}{|k|^{\zeta_2}} \sin^2(\pi k \cdot (x-y)) \\ &+ \frac{4}{C^2} \sum_{k \in \mathbb{Z}^3 \setminus \{0\}} \mathrm{d}_k{}^2 \frac{(1 - \mathrm{e}^{-\lambda_k t})^2}{|k|^{\zeta_2}} \sin^2(\pi k \cdot (x-y)), \end{aligned}$$

again by using the lower estimate in (2.32). As $t \to \infty$, we get

$$S_2(x,y,\infty) \sim \frac{4\pi^{\zeta_2}}{C^2} \sum_{k \in \mathbb{Z}^3 \setminus \{0\}} \left[\mathrm{d}_k{}^2 + \left(\frac{C}{2}\right) c_k \right] |x-y|^{\zeta_2},$$

when $|x-y|$ is small, where $\zeta_2 = 2/3 + \tau_2 \approx 0.696$.

Similarly

$$
\begin{aligned}
S_3 &= E(|u(x,t)-u(y,t)|^3) \\
&\le \frac{2^3}{C^3} \sum_{k\in\mathbb{Z}^3\backslash\{0\}} \frac{[|\mathrm{d}_k|^3(1-\mathrm{e}^{-\lambda_k t})^3 + 3(C/2)c_k|\mathrm{d}_k|(1-\mathrm{e}^{-2\lambda_k t})(1-\mathrm{e}^{-\lambda_k t})]}{|k|} \\
&\quad \times|\sin^3(\pi k\cdot(x-y))|,
\end{aligned}
$$

and

$$S_3(x,y,\infty) \sim \frac{2^3\pi}{C^3} \sum_{k\in\mathbb{Z}^3\backslash\{0\}} [|\mathrm{d}_k|^3 + 3(C/2)c_k|\mathrm{d}_k|]|x-y|,$$

where $\tau_3 = 0$.

All the structure functions are computed in a similar manner; for the pth structure functions, we get that S_p is estimated by

$$S_p \le \frac{2^p}{C^p} \sum_{k\in\mathbb{Z}^3\backslash\{0\}} \frac{\sigma^p\cdot(-\mathrm{i}\sqrt{2})^p\, U\left(-\frac{1}{2}p, \frac{1}{2}, -\frac{1}{2}(M/\sigma)^2\right)}{|k|^{\zeta_p}} |\sin^p(\pi k\cdot(x-y))|,$$

where U is the confluent hypergeometric function, $M = |\mathrm{d}_k|(1-\mathrm{e}^{-\lambda_k t})$, and $\sigma = \sqrt{(C/2)c_k(1-\mathrm{e}^{-2\lambda_k t})}$. Thus the coefficients of S_p are given by the raw moments of a Gaussian, the first few of which are listed in Table 2.1. Now $S_p(x,y,\infty)$ is

Table 2.1 Moments of a Gaussian

Order	Raw moment	Central moment	Cumulant
1	M	0	M
2	$M^2 + \sigma^2$	σ^2	σ^2
3	$M^3 + 3M\sigma^2$	0	0
4	$M^4 + 6M^2\sigma^2 + 3\sigma^4$	$3\sigma^4$	0
5	$M^5 + 10M^3\sigma^2 + 15M\sigma^4$	0	0
6	$M^6 + 15M^4\sigma^2 + 45M^2\sigma^4 + 15\sigma^6$	$15\sigma^6$	0
7	$M^7 + 21M^5\sigma^2 + 105M^3\sigma^4 + 105M\sigma^6$	0	0
8	$M^8 + 28M^6\sigma^2 + 210M^4\sigma^4 + 420M^2\sigma^6 + 105\,\sigma^8$	$105\sigma^8$	0

$$S_p \sim \frac{2^p\pi^{\zeta_p}}{C^p} \sum_{k\in\mathbb{Z}^3\backslash\{0\}} ((C/2)c_k)^{p/2}\cdot(-i\sqrt{2})^p\, U\left(-\frac{1}{2}p, \frac{1}{2}, -\frac{\mathrm{d}_k^2}{Cc_k}\right)|x-y|^{\zeta_p},$$

to leading order for $|x-y|$ small. We also obtain Kolmogorov's $4/5$ law, see [26],

$$S_3 = -\frac{4}{5}\varepsilon(0)|x-y|,$$

to leading order, where ε is the mean energy dissipation rate (1.8).

2.8 The Solution of the Stochastic Linearized Navier–Stokes Equation

In this section we will study the Navier–Stokes equation (1.1) linearized but with the same noise as the stochastic Navier–Stokes equation (1.66). This will allow us to use some of the methods already discussed in this chapter and write the following initial value problem as an integral equation. The operator K plays a central role again and the Feynman–Kac formula, Lemma 2.3, from Sect. 2.2.1 utilized. We will also find the martingale that removes the transport term from the equation. The linearized Navier–Stokes equation with fully developed turbulent noise is

$$\begin{aligned} dz &= (\nu\Delta z - u\cdot\nabla z - z\cdot\nabla u + 2\nabla\Delta^{-1}\mathrm{tr}(\nabla u\nabla z))dt + \sum_{k\neq 0} c_k^{\frac{1}{2}}\, db_t^k e_k(x) \\ &\quad + \sum_{k\neq 0} d_k|k|^{1/3} dt e_k(x) + z\sum_{k\neq 0}^{M}\int_{\mathbb{R}} h_k\bar{N}^k(dt,dz), \\ z(x,0) &= z_0(x). \end{aligned} \tag{2.38}$$

In distinction to the Navier–Stokes equations (1.66), (2.38) can be solved by use the Feynman–Kac formula, Lemma 2.3, and Cameron–Martin (or Girsanov's Theorem 2.2). The solution can be written as

$$\begin{aligned} z &= e^{Kt}e^{\int_0^t dq}M_t z^0 + \sum_{k\neq 0} c_k^{1/2}\int_0^t e^{K(t-s)}e^{\int_s^t dq}M_{t-s}\, db_s^k e_k(x) \\ &\quad + \sum_{k\neq 0} d_k\int_0^t e^{K(t-s)}e^{\int_s^t dq}M_{t-s}|k|^{1/3}ds\; e_k(x), \end{aligned} \tag{2.39}$$

where K is the operator

$$K = \nu\Delta + 2\nabla\Delta^{-1}\mathrm{tr}(\nabla u\nabla),$$

(that was called $\bar{K}$ above) and

$$\begin{aligned} 3\int_s^t dq &= \sum_{k\neq 0}^{M}\left\{\int_0^t\int_{\mathbb{R}} \ln(1+h_k)\bar{N}^k(ds,dz) + \int_0^t\int_{\mathbb{R}}(\ln(1+h_k)-h_k)m_k(ds,dz)\right\} \\ &= \ln(|k|^{2/3}(2/3)^{N_t^k}), \end{aligned}$$

by Ito's formula and a computation similar to the one that produces the geometric Lévy process, see Sect. 2.3, $m_k(dt,dz)$ being the kth Lévy measure. As in Sects. 2.4 and 2.5, M_t is the martingale:

$$M_t = \exp\{-\int_0^t u(B_s,s)\cdot dB_s - \frac{1}{2}\int_0^t |u(B_s,s)|^2 ds\}. \tag{2.40}$$

It is the Radon–Nikodym derivative of the measure, see Theorem 2.2, of the associated Ito process

$$dx_t = -u dt + \sqrt{2\nu} dB_t,$$

u being the fluid velocity and $B_t = (B_t^1, B_t^2, B_t^3)$ a three-dimensional vector of auxiliary Brownian motions, with respect to the usual Brownian measure.

In the next chapter we will use the stochastic linearized Navier–Stokes equation to construct the invariant measure of turbulence.

Chapter 3
The Invariant Measure and the Probability Density Function

3.1 The Invariant Measure of the Stochastic Navier–Stokes Equation

The invariant measure of the stochastic Navier–Stokes equation determines all the one-point statistics of turbulence, or the statistics of quantities defined at one point x in the flow. This quantity determines all the statistical properties of the turbulent velocity field, see [56], and in distinction to the nonlinear Navier–Stokes equation, the invariant measure satisfies a linear but a functional differential equation; see [56]. In fact Hopf [29] found a linear equation for the characteristic function (Fourier transform) of the invariant measure in 1952, but at that time methods for solving such an equation were not available. In Hopf's equation the noise for fully developed turbulence was missing, but in Kolmogorov's equation for the invariant measure the noise is always supplied. Since only the linearized Navier–Stokes equation (2.38) appears below in the Kolmogorov–Hopf equation for the invariant measure, we will think about the linearized Navier–Stokes equation as the infinite-dimensional Ito process, whose generator gives the Kolmogorov–Hopf equation. Thus associated with such an Ito process is a diffusion equation, a linear functional differential equation, that is the Kolmogorov–Hopf equation determining the invariant measure. We will now derive this equation. This will make clear how to compute the coefficients in the Kolmogorov–Hopf equation.

The Kolmogorov–Hopf equation for the invariant measure is

$$\frac{\partial \phi}{\partial t} = \frac{1}{2}\mathrm{tr}[P_t C P_t^* \Delta\phi] + \mathrm{tr}[P_t \bar{D} \nabla \phi] + \langle \bar{K}(z) P_t, \nabla\phi\rangle, \tag{3.1}$$

where $\bar{D} = (|k|^{1/3} d_k)$, $\phi(z)$ is a bounded function of z and $|x| = \langle x, x\rangle^{1/2}$ where $\langle \cdot,\cdot\rangle$ is the inner product on H. Here $C^{1/2}$, $D \in L(H)$ are linear operators on $H = L^2(\mathbb{T}^3)$, defined by

$$C^{1/2}u = \sum_{k\neq 0} C_k^{\frac{1}{2}} \hat{u}_k e_k, \;\; Du = \sum_{k\neq 0} D_k \hat{u}_k e_k$$

B. Birnir, *The Kolmogorov-Obukhov Theory of Turbulence: A Mathematical Theory of Turbulence*, SpringerBriefs in Mathematics, DOI 10.1007/978-1-4614-6262-0_3,

for $u = \sum_{k\neq 0} \hat{u}_k e_k \in L^2(\mathbb{T}^3)$, $C_k^{1/2}$ and D_k are 3-by-3 diagonal matrices with entries $c_{k,j}^{1/2}$ and $\mathrm{d}_{k,j}$, $j = 1,2,3$ on the diagonal:

$$P_t = \mathrm{e}^{-\int_0^t \nabla u \, \mathrm{d}r} \prod_k^m (|k|^{2/3}(2/3)^{N_t^k})^{\frac{1}{3}},$$

by the computation of how the log-Poisson processes are produced, from the stochastic Navier–Stokes equation, by the Feynman–Kac formula (2.27) above. The operator $\bar{K}$ is the linearized Navier–Stokes operator

$$\bar{K} = \nu\Delta - u \cdot \nabla + 2\nabla\Delta^{-1}\mathrm{tr}(\nabla u \nabla) = K - u \cdot \nabla,$$

and z is the solution of the linearized Navier–Stokes equation (2.38). Notice that now K has 2 in front of the pressure term as in Sect. 2.8.

To find the infinite-dimensional Ito process whose Kolmogrov's backward equation is (3.1), we consider the linearized Navier–Stokes equation with the same noise as (1.65); see Sect. 2.8. This is the functional derivative of the deterministic Navier–Stokes equation (1.1), driven with the same noise as the stochastic equation (1.65), to give an Ito process in function space. It is analogous to the stochastic evolution of the volume element in finite dimensions, but here the Ito process determines the evolution of any bounded function of u, in infinite dimensions; see [56]. The solution of the linearized Navier–Stokes equation (2.38) can be written in integral form as

$$z = \mathrm{e}^{Kt} P_t M_t z^0 + \sum_{k\neq 0} c_k^{1/2} \int_0^t \mathrm{e}^{K(t-s)} P_{t-s} M_{t-s} \mathrm{d}b_s^k e_k(x)$$
$$+ \sum_{k\neq 0} \mathrm{d}_k \int_0^t \mathrm{e}^{K(t-s)} P_{t-s} M_{t-s} |k|^{1/3} \mathrm{d}s \, e_k(x) \tag{3.2}$$

by the Feynman–Kac formula, where the operator K generates the flow e^{Kt} and

$$M_t = \exp\left\{ -\int u(B_s,s)\mathrm{d}B_s - \frac{1}{2}\int_0^t |u(B_s,s)|^2 \mathrm{d}s \right\}$$

is a martingale with $B_t \in \mathbb{R}^3$ an auxiliary Browninan motion; see Sect. 2.8.

Now we define the variance

$$Q_t = \int_0^t \mathrm{e}^{K(s)} P_s M_s C M_s P_s^* \mathrm{e}^{K^*(s)} \mathrm{d}s \tag{3.3}$$

and drift

$$E_t = \int_0^t \mathrm{e}^{K(s)} P_s M_s \bar{D} \mathrm{d}s \tag{3.4}$$

operators. Then the solution of the Kolmogorov–Hopf equation (3.1) can be written in the form

$$\begin{aligned} R_t\phi(z) &= \int_H \phi(y)\mathcal{N}_{(e^{Kt}P_tM_tz+E_tI,Q_t)} * \mathbb{P}_{P_t}(dy) \\ &= \int_H \phi(e^{Kt}P_tM_tz+E_tI+y)\mathcal{N}_{(0,Q_t)} * \mathbb{P}_{P_t}(dy), \end{aligned}$$

where $\mathbb{P}_{P_t}$ is the Poisson law of P_t. $\mathcal{N}_{m,Q_t}$ denotes the infinite-dimensional normal distribution on H with mean m and variance Q_t; see [56], $I = \sum e_k$ and $E_tI \in H$.

3.1.1 The Invariant Measure of Turbulence

We can now write a formula for the invariant measure of turbulence.

Theorem 3.1. *The invariant measure of the stochastic Navier–Stokes equation on* $H_c = H^{3/2^+}(\mathbb{T}^3)$ *has the form*

$$\mu(dx) = e^{\langle Q^{-1/2}EI,\, Q^{-1/2}x\rangle - \frac{1}{2}|Q^{-1/2}EI|^2}\mathcal{N}_{(0,Q)}(dx)\sum_k \delta_{k,l}\prod_{j\neq l}^{m}\delta_{N_t^j}\sum_{j=0}^{\infty} p_{m_l}^j \delta_{(N_t^l-j)}, \tag{3.5}$$

where $Q = Q_\infty$, $E = E_\infty$, $m_k = \ln|k|^{2/3}$ *is the mean of the log-Poisson processes (2.26) and* $p_{m_k}^j = \frac{(m_k)^j e^{-m_k}}{j!}$ *is the probability of* $N_\infty^k = N_k$ *having exactly* j *jumps, and* $\delta_{k,l}$ *is the Kronecker delta function.*

Suppose that the operator Q is trace-class, $E(Q^{1/2}H) \subset Q^{1/2}(H)$, and that $e^{Kt}P_tM_t(H) \subset Q_t^{1/2}(H)$, $t > 0$, where $H = H_c$, then, with u given, the invariant measure μ is unique, ergodic, and strongly mixing. We know that the above invariant measure is unique for the strong swirl [17] and strong rotation [3, 4], but it depends on u, and its uniqueness for general turbulent flows depends on the uniqueness of u.

The proof of Theorem 3.1 uses the above machinery and is analogous to the proof of Theorem 8.20 in [56].

Proof. The (log) Poissonian part of the invariant measure is the same for all t, so we just have to proof that the infinite-dimensional Gaussian is invariant or that

$$\int_H R_t\phi(y)\mathcal{N}_{(E_\infty x,Q_\infty)}(dy) = \int_H \phi(y)\mathcal{N}_{(E_\infty x,Q_\infty)}(dy),$$

where $E_\infty = E$. We set $P_t^{\text{new}} = P_t^{\text{old}}M_t$ for the ease of writing. Recall that

$$R_t\phi(z) = \int_H \phi(y)\mathcal{N}_{(e^{Kt}P_tz+E_tx,Q_t)}(dy) = \int_H \phi(e^{Kt}P_tz+E_tx+y)\mathcal{N}_{(0,Q_t)}(dy).$$

We let $\phi(z) = e^{i\langle z,h\rangle}$; then

$$\int_H e^{i\langle y,h\rangle}\mathcal{N}_{(E_\infty x,Q_\infty)}(dy) = e^{i\langle E_\infty x,h\rangle - \frac{1}{2}\langle Q_\infty h,h\rangle}$$

by a standard computation and

$$\begin{aligned}\int_H R_t\phi(y)\mathcal{N}_{(E_\infty x,Q_\infty)}(dy) &= \int_H e^{i\langle y,h\rangle}\mathcal{N}_{(e^{Kt}P_t z+E_t x,Q_t)} * \mathcal{N}_{(E_\infty x,Q_\infty)}(dy)\\ &= \int_H e^{i\langle e^{Kt}P_t z+E_t x,h\rangle-\frac{1}{2}\langle Q_t h,h\rangle}\mathcal{N}_{(E_\infty x,Q_\infty)}(dy)\\ &= e^{i\langle E_t x,h\rangle-\frac{1}{2}\langle Q_t h,h\rangle}e^{i\langle E_\infty x,P_t^* e^{K^*t}h\rangle-\frac{1}{2}\langle Q_\infty P_t^* e^{K^*t}h,P_t^* e^{K^*t}h\rangle}\\ &= e^{i\langle E_\infty x,h\rangle-\frac{1}{2}\langle Q_\infty h,h\rangle}\end{aligned}$$

if

$$e^{Kt}P_t Q_t P_t^* e^{K^*t}+Q_t = Q_\infty$$

and

$$e^{Kt}P_t E_\infty + E_t = E_\infty.$$

The last two equations are easily verified using the definitions (3.3) and (3.4), if we interpret

$$e^{Kt}P_t e^{Ks}P_s = e^{K(t+s)}P_{(t+s)},$$

as solving the linearized stochastic Navier–Stokes equation with initial data $e^{Ks}P_s$ for the time interval $[s,t+s]$. The equality hold by existence of the flow; see Lemma 2.6.

Next we show that given u the invariant measure μ is unique. Again it suffices to show this for the infinite-dimensional Gaussian. We consider the Fourier transform of μ; by the invariance of μ

$$\hat{\mu}(P_t^* e^{K^*t}h)e^{i\langle E_t x,h\rangle-\frac{1}{2}\langle Q_t h,h\rangle} = \hat{\mu}(h)$$

as $t\to\infty$, we get

$$\hat{\mu}(h) = e^{i\langle E_\infty x,h\rangle-\frac{1}{2}\langle Q_\infty h,h\rangle},$$

since $P_t^* e^{K^*t}h\to 0$. Now E_∞ and Q_∞ depend on u but given u the Fourier transform is unique and thus

$$\mu = \mathcal{N}_{(E_\infty x,Q_\infty)}.$$

Finally we show that

$$\lim_{t\to\infty} R_t\phi(z) = \int_H \phi(y)\mathcal{N}_{(E_\infty z,Q_\infty)}(dy)$$

in $L^2(H,\mu)$, $z\in H$. This implies that μ is ergodic and strongly mixing, since the Poissonian part of the measure is the same for all t. We consider $\phi(z)=e^{i\langle z,h\rangle}$ again:

$$\begin{aligned}\lim_{t\to\infty} R_t\phi &= \lim_{t\to\infty} e^{i\langle E_t x,h\rangle-\frac{1}{2}\langle Q_t h,h\rangle}e^{\langle z,P_t^* e^{K^*t}h\rangle}\\ &= e^{i\langle E_\infty x,h\rangle-\frac{1}{2}\langle Q_\infty h,h\rangle} = \hat{\mu}(h).\end{aligned}$$

The conclusion follows by the uniqueness of the Fourier transform.

3.2 The Invariant Measure for the Velocity Differences

We will now find the Kolmogorov–Hopf functional differential equation for the invariant measure of the Navier–Stokes equation for the velocity differences:

$$z = u - w = u(x,t) - u(y,t).$$

The previous measure was the measure determining the one-point statistics, but the measure for the velocity difference will determine the two-point statistics. We are simplifying this a little using isotrophy; namely, in general, the velocity difference is a tensor. The linearized Navier–Stokes operator is now

$$\bar{K} = \nu\Delta - u\cdot\nabla + \nabla\Delta^{-1}\mathrm{tr}((\nabla u + \nabla w)\nabla),$$

but otherwise the derivation is similar to the derivation of the 1-point measure above. The formula for the 2-point measure is the same (3.5), but now the operator K depends on the two points x and y and therefore the variance (3.3) and the drift (3.4) will also depend on these two points. In fact the measure depends on the lag variable $x-y$. A better way of capturing the dependence on the lag variable is to write the difference of the inertial terms as

$$-u\cdot\nabla w + w\cdot\nabla u = -u\cdot\nabla(u-w) - (u-w)\cdot\nabla u + (u-w)\cdot\nabla(u-w).$$

This produces the new operator

$$\tilde{K} = \nu\Delta - u\cdot\nabla + z\cdot\nabla - \nabla u + \nabla\Delta^{-1}\mathrm{tr}((\nabla u + \nabla w)\nabla) = K - u\cdot\nabla + z\cdot\nabla - \nabla u$$

with the understanding that now K is a function of $\left(\frac{(u+w)}{2}\right)$ through the pressure term. The last three terms are removed by a combination of Feynman–Kac and the Cameron-Martin formula (Girsanov's theorem) and we get the martingale

$$M_t = \exp\left\{\int_0^t u(x - B_{-s} + y, s)\cdot \mathrm{d}B_{-s} + \int_0^t z(B_s)\cdot \mathrm{d}B_s \right.$$
$$\left. -\frac{1}{2}\int_0^t |u(x - B_{-s} + y, s) + z(B_s), s)|^2 \mathrm{d}s\right\}$$

after a time reversal of the auxiliary Brownian motion B_t; see [44]. The computation of the measure follows the procedure for the computation of the measure for the 1-point statistics. The difference of the two equations (for u and w) is written as an integral equation

$$z = \mathrm{e}^{K(t)}\mathrm{e}^{-\int_0^t \nabla u\,\mathrm{d}s}\mathrm{e}^{\int_0^t \mathrm{d}q} M_t z^0 + \sum_{k\neq 0} c_k^{1/2}\int_0^t \mathrm{e}^{K(t-s)}\mathrm{e}^{-\int_s^t \nabla u\,\mathrm{d}r}\mathrm{e}^{\int_s^t \mathrm{d}q} M_{t-s}\mathrm{d}b_s^k e_k(x)$$
$$+ \sum_{k\neq 0} \mathrm{d}_k \int_0^t \mathrm{e}^{K(t-s)}\mathrm{e}^{-\int_s^t \nabla u\,\mathrm{d}r}\mathrm{e}^{\int_s^t \mathrm{d}q} M_{t-s}|k|^{1/3}\mathrm{d}s\, e_k(x) \tag{3.6}$$

by the Feynman–Kac formula and Girsanov's theorem where K is the operator

$$K = \nu\Delta + \nabla\Delta^{-1}\mathrm{tr}((\nabla u + \nabla w)\nabla), \tag{3.7}$$

and

$$P_t = \mathrm{e}^{-\int_0^t \nabla u\, ds} \mathrm{e}^{\int_0^t dq} M_t = \mathrm{e}^{-\int_0^t \nabla u\, dr} \prod_k |k|^{2/3} (2/3)^{N_t^k} M_t.$$

The Kolmogorov–Hopf equation for the Ito processes (3.6) now becomes

$$\frac{\partial \phi}{\partial t} = \frac{1}{2}\mathrm{tr}[P_t C P_t^* \Delta\phi] + \mathrm{tr}[P_t \bar{D} \nabla\phi] + \langle K(z)P_t, \nabla\phi\rangle, \tag{3.8}$$

where $\bar{D} = (|k|^{1/3} D_k)$ and $\phi(z)$ is a bounded function of z. It is also the Kolmogorov backward equation of the Ito process (3.6).

The variance is

$$Q_t = \int_0^t \mathrm{e}^{K(s)} P_s C P_s^* \mathrm{e}^{K^*(s)} ds \tag{3.9}$$

and the drift is

$$E_t = \int_0^t \mathrm{e}^{K(s)} P_s \bar{D} ds. \tag{3.10}$$

Then the solution of the Kolmogorov–Hopf equation (3.8) can be written in the form

$$\begin{aligned} R_t \phi(z) &= \int_H \phi(y) \mathcal{N}_{(\mathrm{e}^{K(t)} P_t z + E_t I, Q_t)} * \mathcal{N}_{(0,2\nu)} * \mathbb{P}_{P_t}(dy) \\ &= \int_H \phi(\mathrm{e}^{K(t)} P_t z + E_t I + y) \mathcal{N}_{(0,Q_t)} * \mathcal{N}_{(0,2\nu)} * \mathbb{P}_{P_t}(dy), \end{aligned} \tag{3.11}$$

where $\mathbb{P}_{P_t}$ is the Poisson law of P_t; see [56]. Here $|x| = \langle x, x\rangle^{1/2}$ where $\langle\cdot,\cdot\rangle$ is the inner product on H and $z = z_0$. $\mathcal{N}_{m,Q_t}$ denotes the infinite-dimensional normal distribution on H with mean m and variance Q_t, $I = \sum e_k, E_t I \in H$ and $\mathcal{N}_{(0,2\nu)}$ the law of the three-dimensional Brownian motion in the martingale M_t. If Q_t is of trace-class $Q_t \in L^+(H)$, then R_t is Markovian.

Theorem 3.2. *The invariant measure for the velocity differences (two-point statistics) of the Navier–Stokes equation on $H_c = H^{3/2^+}(\mathbb{T}^3)$ has the form*

$$\begin{aligned} \mu(dx, dy) &= \mathrm{e}^{\langle Q^{-1/2}EI,\, Q^{-1/2}x\rangle - \frac{1}{2}|Q^{-1/2}EI|^2} \mathcal{N}_{(0,Q)}(dx) \\ &\quad * \mathcal{N}_{(0,2\nu)}(dy) \sum_k \delta_{k,l} \sum_{j=0}^{\infty} p_{m_l}^j \delta_{(N_l - j)}, \end{aligned} \tag{3.12}$$

where $Q = Q_\infty$, $E = E_\infty$. Here $m_k = \ln|k|^{2/3}$ is the mean of the log-Poisson processes (2.26) and $p_{m_k}^j = \frac{(m_k)^j \mathrm{e}^{-m_k}}{j!}$ is the the probability of $N_\infty^k = N_k$ having exactly j jumps, and $\delta_{k,l}$ is the Kronecker delta function.

Suppose that the operator Q is trace-class, $E(Q^{1/2}H) \subset Q^{1/2}(H)$, and that $e^{K(t)}P_tM_t(H) \subset Q_t^{1/2}(H)$, $t > 0$, where $H = L^2(\mathbb{T}^3)$, then, given u, the invariant measure μ is unique, ergodic, and strongly mixing. The proof of Theorem 3.2 is similar to the proof of Theorem 3.1.

It is easy to check that the moments of the invariant measure for the two-point statistics give the estimates for the structure functions above. The variable in the latter three-dimensional Gaussian $\mathcal{N}_{(0,2\nu)}(dy)$ in the invariant measure is the lag variable.

The same comments as above apply to the measure (3.12) as the invariant measure for the one-point statistics (3.5). It is unique for the strong swirl [17] and strong rotation [3, 4], but its uniqueness for general turbulent flows depends on the uniqueness of u.

3.3 The Differential Equation for the Probability Density Function

We must compute the probability density function (PDF) of the invariant measure (3.5), for the velocity differences, in order to compare with PDFs constructed from simulations and experiments. The simplest way of doing this is to derive the differential equation for the distribution function from the Kolmogorov–Hopf equation (3.1). We start by rewriting the equation Kolmogorov–Hopf (3.1) in the form

$$\frac{\partial \phi}{\partial t} = \frac{1}{2}\text{tr}[Q_t \Delta \phi] + \text{tr}[E_t \nabla \phi], \tag{3.13}$$

where Q_t and E_t are respectively the variance (3.9) and drift (3.10), but computed with the operator K in (3.7). This can be done by redefining the underlying infinite-dimensional Ito process with the formulas (3.9) and (3.10) for the variance and the drift. We have to take the trace of the functional variables to get the equation for the PDF. The resulting equation is

$$\frac{\partial \phi}{\partial t} = \frac{1}{2}\Delta \phi + \frac{1}{\sqrt{2t}} c \cdot \nabla \phi, \tag{3.14}$$

where $\hat{c}(|k|) = (Q_t^{-1/2}E_t)_k$ are Fourier coefficients of c, after we scale by the variance Q_t. Now scaling the equation by $-2t$ and sending $t \to \infty$ gives the equation

$$\frac{1}{2}\Delta \phi + c \cdot \nabla \phi = \phi, \tag{3.15}$$

with a trivial rescaling of time. This is the (stationary) equation for the distribution function. Now the PDF is for the absolute value of the velocity differences $w = |u(x,t) - u(y,t)|$, by (3.20) below, so the angle derivatives of w do not appear, and $\hat{c} = (Q^{-1/2}E)_k \sim \bar{c}|k|^{1/3}/|k|^{1/3} = \bar{c}$ for k large. (The intermittency corrections wash out.) Thus, taking the trace of the spatial (lag) variables also, we get that $c = \frac{\bar{c}}{w}$. In polar cordinates $\Delta\phi = \phi_{ww} + \frac{2}{w}\phi_w$, in three dimensions. Thus (3.15) becomes

$$\frac{1}{2}\phi_{ww} + \frac{1+\bar{c}}{w}\phi_w = \phi. \tag{3.16}$$

This is the stationary equation satisfied by the PDF.

Example 3.1. The above computation is clarified by the following example. Consider the equation

$$\phi_t = \phi_{xx} + \frac{c}{\sqrt{2t}}\phi_x,$$

where $\phi = \frac{e^{-(x-a)^2/b}}{\sqrt{\pi b}}$ is a Gaussian. It is easy to check that this equation holds if $a_t = -\frac{c}{\sqrt{2t}}$ and $b_t = 4$, so $a = -c\sqrt{2t}$ and $b = 4t$. Thus invariant measure is produced by scaling out t:

$$\phi(y)\mathrm{d}y = \frac{e^{-\frac{(y+c)^2}{2}}}{\sqrt{2\pi}}\mathrm{d}y = \frac{e^{\frac{(y-\frac{a}{\sqrt{b/2}})^2}{2}}}{\sqrt{2\pi}}\mathrm{d}y = \phi(x,t)\mathrm{d}x,$$

where $y = x/\sqrt{2t}$. This invariant measure satisfies the stationary equation (3.15).

3.4 The PDF for the Turbulent Velocity Differences

It is now possible to compute the PDF for the velocity differences in turbulence. The form of (3.16) suggests that we should look for a solution of the form $f = x^a K_\lambda$ where K_λ is a modified Bessel's function of the second kind, satisfying the equation

$$K_{xx} + \frac{1}{x}K_x - \left(1 + \frac{\lambda^2}{x^2}\right)K = 0.$$

A substitution of this ansatz into (3.16) gives $a = -\bar{c}$ and $\lambda = \sqrt{\frac{\bar{c}(\bar{c}+1)}{2}}$. The solution is the generalized hyperbolic distribution; see Barndorff-Nielsen [6] and Appendix C. It has an algebraic cusp at the origin and exponential tails and is constructed by multiplying the modified Bessel's function of the second kind K_λ, by $x^{-\lambda}$. For the zeroth moment we get a distinguished solution $\lambda = \bar{c} = 1$ which give the normalized inverse Gaussian (NIG) distribution that was also investigated by Barndorff-Nielsen [7] and used by Barndorff-Nielsen, Blæsild, and Schmiegel to model PDF of velocity increments for several data sets in [8]. It turns out that the distribution functions for all of the moments can be expressed by the NIG distribution function. However, since the intermittency corrections are different for the different moments, the NIG distributions for the different moments have different parameters, as will be explained below.

The PDF of the NIG is

$$\frac{\alpha\delta K_1\left(\alpha\sqrt{\delta^2+(x-\mu)^2}\right)}{\pi\sqrt{\delta^2+(x-\mu)^2}}\,\mathrm{e}^{\delta\gamma+\beta(x-\mu)}. \tag{3.17}$$

The parameters are:

$$\begin{aligned}&\alpha \text{ heavyness of the tail, } \beta \quad \text{asymmetry, } \delta \quad \text{scaling}\\ &\mu \text{ centering, and } \gamma=\sqrt{\alpha^2-\beta^2}.\end{aligned}$$

The NIG distribution has very nice properties that are summarized in [8]. In particular its characteristic function and all of its moments are easily computed. However, the moments of the velocity differences are not the moments of the same NIG distributions, because of the intermittency correction. In fact, the invariant measure (3.12) has both a continuous and a discrete part and because of this each moment comes with its own PDF, as mentioned above. All of these PDFs are solutions of the stationary equation (3.16) and they can be expressed in terms of NIG distributions. However, their parameters α,β,δ, and μ all depend on the particular moment for which one is computing the PDF. Thus these parameters are different for the different moments. The cumulant-generating function $\mu z+\delta(\gamma-\sqrt{\alpha^2-(\beta+z)^2})$ is particularly simple for the NIG and this makes the moments easy to compute; see [8]. The first few moments and the characteristic function of the NIG distribution are:

$$\begin{array}{cl}\text{Mean} & \mu+\delta\beta/\gamma\\ \text{Variance} & \delta\alpha^2/\gamma^3\\ \text{Skewness} & 3\beta/(\alpha\sqrt{\delta\gamma})\\ \text{Excess kurtosis or flatness} & 3(1+4\beta^2/\alpha^2)/(\delta\gamma)\\ \text{Characteristic function} & \mathrm{e}^{\mathrm{i}\mu z+\delta(\gamma-\sqrt{\alpha^2-(\beta+\mathrm{i}z)^2})}.\end{array} \tag{3.18}$$

However, since the parameters α,β,δ, and μ are different for different moments, care must be taken when the moments above are used to compute these parameters. This will be discussed in more details in the next section.

Thus we see that the PDF of the velocity increment is a normalized inverse Gaussian (NIG) distribution that is a generalized hyperbolic distributions with index 1. Using the invariances of the NIG it is given by the four-parameter formula

$$f_j(x,\alpha,\beta,\delta,\mu)=\frac{\alpha\delta\,\mathrm{e}^{\delta\gamma}K_1\left(\alpha\sqrt{\delta^2+(x-\mu)^2}\right)}{\pi\sqrt{\delta^2+(x-\mu)^2}}\,\mathrm{e}^{\beta(x-\mu)},\quad j=1,2, \tag{3.19}$$

where α measures how heavy the exponential part of the tail of the distribution is, β measures how skew the distribution is, δ is a scaling parameter and μ determines

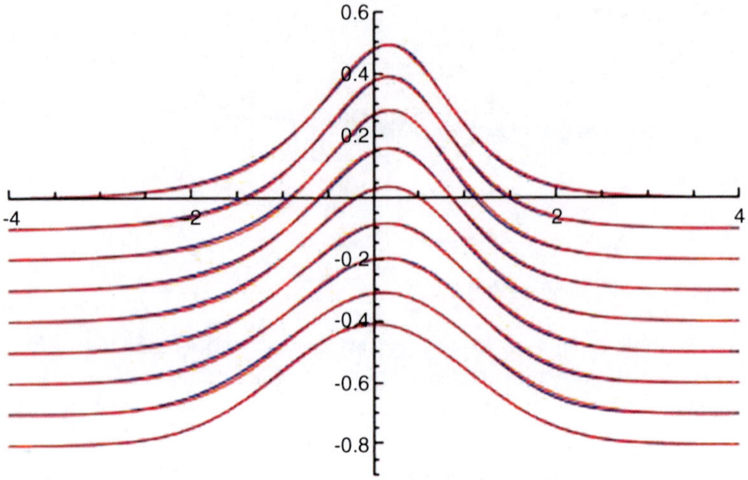

Fig. 3.1 The PDFs from simulations and fits for the longitudinal direction. The PDFs for increasing values of the lag variable are displaced downward. The last PDF looks distinctly Gaussian.

the location (center) of the distribution, $\gamma = \sqrt{\alpha^2 - \beta^2}$. K_1 is the modified Bessel's function of the second kind with index 1. Now the 1st moment of the velocity differences is

$$\begin{aligned} E(\delta_j u) &= E([u(x+s,\cdot) - u(x,\cdot)] \cdot r) = E(|u(x+s,\cdot) - u(x,\cdot)||r|\cos(\theta)) \\ &= \int_{\infty}^{\infty} (xf_j)(x,\alpha,\beta,\delta,\mu)\mathrm{d}x, \end{aligned} \tag{3.20}$$

where $j = 1$, if $r = \hat{s}$ is the longitudinal direction (that is the direction along the lag vector s), and $j = 2$, if $r = \hat{t}$ where $t \perp s$ is a transversal direction and $\hat{r}$ and $\hat{t}$ being unit vectors. θ is the angle between the vectors $[u(x+s,\cdot) - u(x,\cdot)]$ and r, and the absolute value of the former is the reason why the angle derivatives wash out in (3.16). The PDF is symmetric in the transversal direction; then $\beta = \mu = 0$. In that case there are only two independent adjustable parameters, α is the exponential decay at $x = \pm\infty$ and δ is the "peakedness" at the origin. In the nonsymmetric case, there are two more independent adjustable parameters, the skewness parameter β and the centering parameter μ.

The PDF for the velocity increments has the asymptotics,

$$f_j \sim \frac{\delta e^{\delta\gamma}}{\pi} \frac{e^{\beta(x-\mu)}}{(\delta^2 + (x-\mu)^2)}$$

for $(x-\mu)$ small. This is the algebraic (rational) cusp at the origin. The exponential tails are

$$f_j \sim \frac{\sqrt{2}\delta\alpha e^{\delta\gamma - \beta\mu}}{\pi^{3/2}} \frac{e^{-\alpha|x|+\beta x}}{|x|^{3/2}}$$

for $|x|$ large.

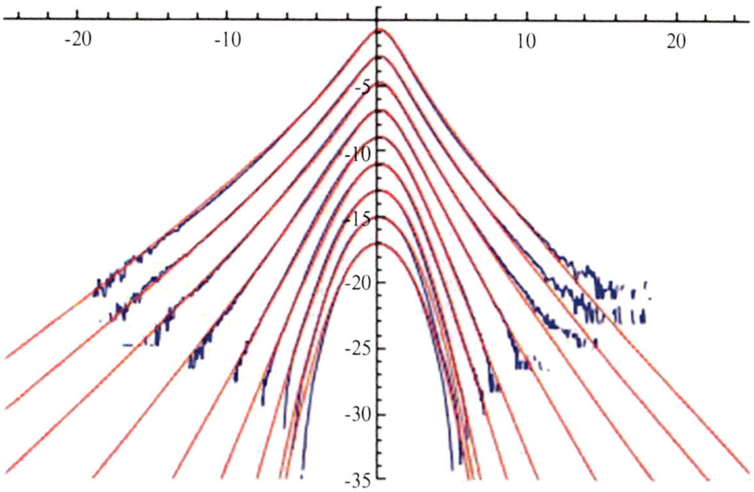

Fig. 3.2 The log of the PDFs from simulations and fits for the longitudinal direction; compare Fig. 4.5 in [76]. Again the logs of PDFs for increasing values of the lag variable are displaced downward. The last ones look Gaussian.

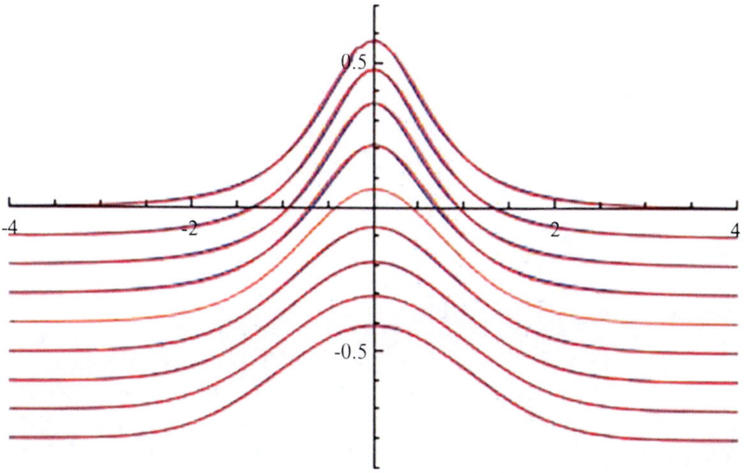

Fig. 3.3 The PDFs from simulations and fits for the transversal direction. The PDFs for increasing values of the lag variable are displaced downward. The last PDF looks distinctly Gaussian.

The exponential tails of the PDF are caused by occasional sharp velocity gradients (rounded-off shocks), whereas the cusp at the origin is caused by the random and gentile fluid motion in the center of the ramps leading up to the sharp velocity gradients; see [40].

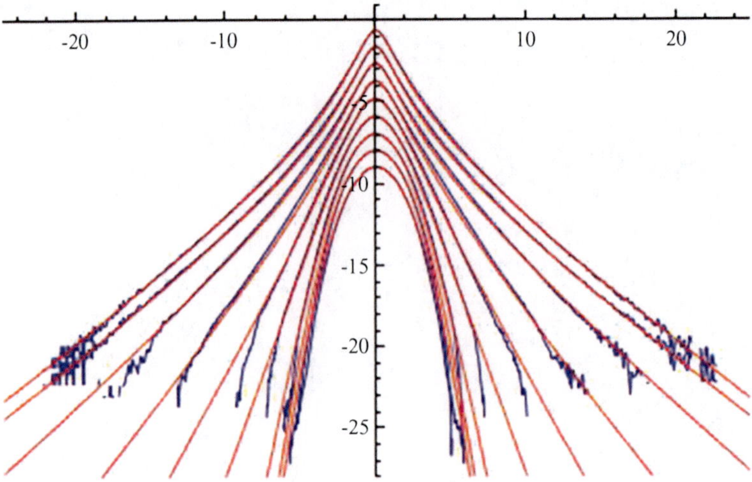

Fig. 3.4 The log of the PDFs from simulations and fits for the transversal direction; compare Fig. 4.6 in [76]. Again the logs of PDFs for increasing values of the lag variable are displaced downward. The last ones look Gaussian and all of them are symmetric and centered at 0.

For large values of the lag variable, the NIG distribution must also approximate a Gaussian. It turns out to do just that. Letting $\alpha, \delta \to \infty$, in the formulas for $f_j(x)$ above, in such a way that $\delta/\alpha \to \sigma$, we get that

$$f_j \to \frac{e^{-\frac{(x-\mu)^2}{2\sigma}}}{\sqrt{2\pi\sigma}} e^{\beta(x-\mu)}.$$

3.5 Comparison with Simulations and Experiments

We now compare the above PDFs with the PDFs found in simulations and experiments, using the first moment $g_j(x) = (xf_j)(x, \alpha, \beta, \delta, \mu)$, where f_j, $j = 1, 2$ are the PDFs in formula (3.19). Because of the discrete jump measure (2.37) all the higher moments come with their own PDF. The PDF for the pth moment is given by the formula

$$f^p_{j\ (\alpha,\beta,\delta,\mu)(p)}(x) = \frac{\alpha\delta\, e^{\delta\gamma} K_1\left(\alpha\sqrt{\delta^2 + (x-\mu)^2}\right)}{\pi\sqrt{\delta^2 + (x-\mu)^2}}\, e^{\beta(x-\mu)}, \tag{3.21}$$

where $\gamma = \sqrt{\alpha^2 - \beta^2}$ and K_1 is the modified Bessel's function of the second kind with index 1, similar to (3.19). The density of the pth moment itself is

$$x^p f^p_{j\ (\alpha,\beta,\delta,\mu)(p)}(x) = \frac{\alpha^{1-p}\delta\, e^{\delta\gamma} K_1\left(\alpha\sqrt{\delta^2 + (x-\mu)^2}\right)}{\pi(\delta^2 + (x-\mu)^2)^{(1-p)/2}}\, e^{\beta(x-\mu)}, \tag{3.22}$$

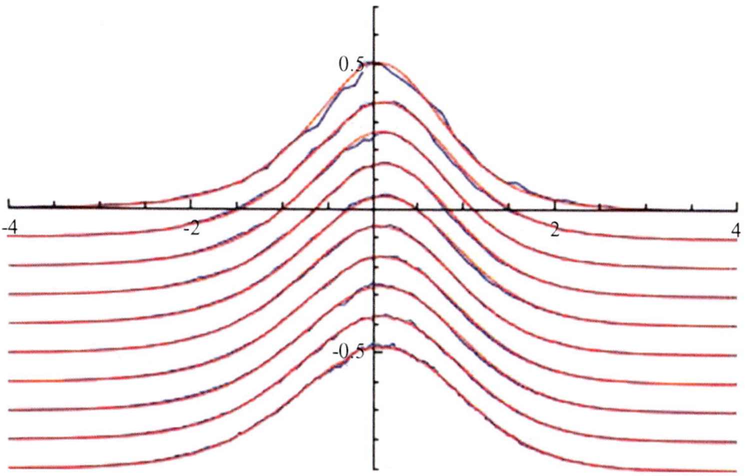

Fig. 3.5 The PDFs from experiments and fits. The PDFs for increasing values of the lag variable are displaced downward.

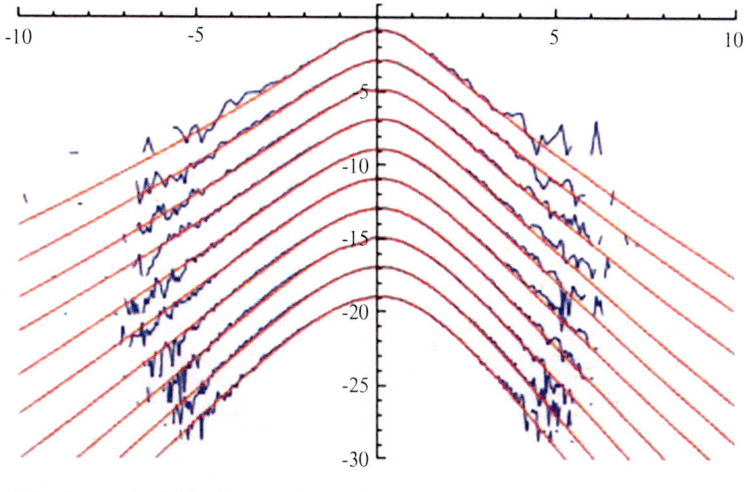

Fig. 3.6 The log of the PDFs from experiments and fits. Again the logs of PDFs for increasing values of the lag variable are displaced downward.

where $j = 1$, for the longitudinal, and $j = 2$ for the transverse component, as in (3.19). All the four parameters α, β, δ, and μ are functions of p because of intermittency.

If the first four moments in (3.18) are given, then the four parameters in the NIG distribution can be computed directly. However, this is probably not the best way to do so. Firstly, this would only give the parameters for the first four moments and the parameters for the higher moments would have to be computed separately. Secondly, since both the longitudinal and the transverse moments can be measured, see Formula (3.20), giving the first four moments may overdetermine the

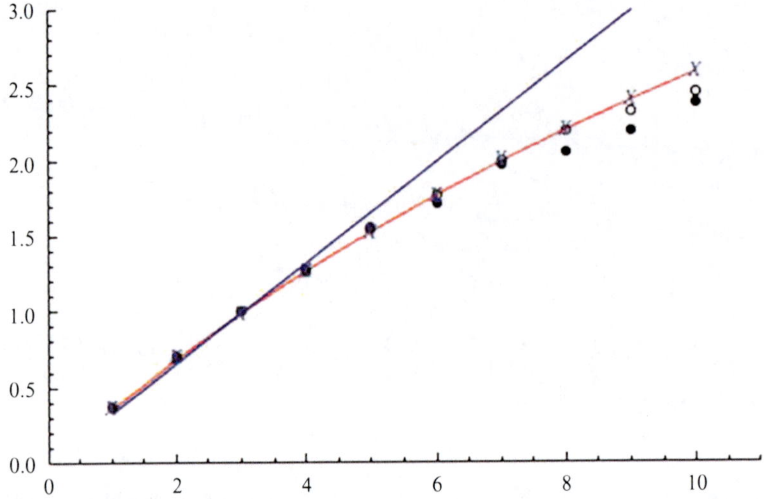

Fig. 3.7 The exponents of the structure functions as a function of order, theory, or Kolmogorov–Obukhov–She–Leveque scaling (*red*), experiments (*disks*), dns simulations (*circles*), from [19], and experiments (X), from [64]. The Kolmogorov–Obukhov '41 scaling is also shown as a *blue line* for comparison.

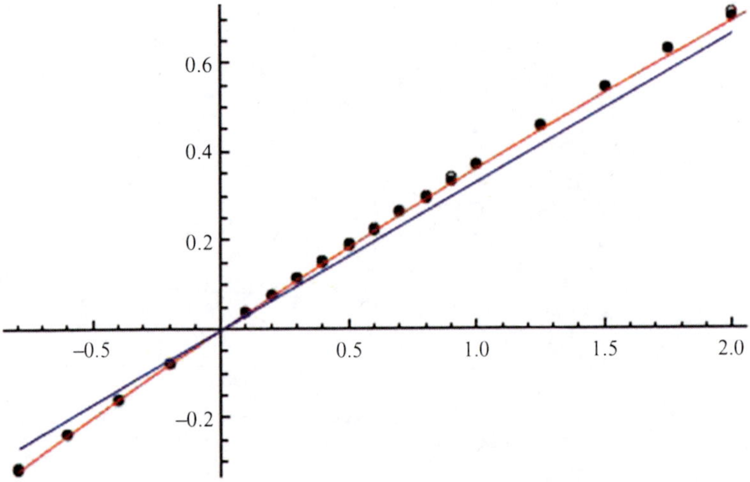

Fig. 3.8 The exponents of the structure functions as a function of order $(-1,2]$, theory, or Kolmogorov–Obukhov–She–Leveque scaling (*red*), experiments (*disks*), and dns simulations (*circles*), from [19]. The Kolmogorov–Obukhov '41 scaling is also shown as a *blue line* for comparison.

four parameters in NIG. A better method is to give both the longitudinal and transverse measurements for two moments. This will determine the four parameters in NIG and give the NIG for these two moments. One is actually giving the NIG of the projection onto these two moments in moments space. From a theoretical point of view it makes sense to always give the measurements for the third moment, because it does not have any intermittency corrections, corresponding to Kolmogorov's $4/5$ law. Thus one can say given the longitudinal and transverse measurements for the third moment the PDF (NIG) for every moment is determined by the longitudinal and transverse measurements for that moment. However, it may depend on the experiment whether this is the most practical projection.

The direct Navier–Stokes simulations (DNS), in Figs. 3.1–3.4, were provided by Michael Wilczek from his Ph.D. thesis; see [76]. The simulations are plotted in blue and the fits in red. The experimental results in Figs. 3.5 and 3.6 are from the particle tracking experiments by Eberhard Bodenschatz's group. The PDFs of Eulerian velocity differences are obtained from the instantaneous particle velocities by conditioning on given spatial separations; see [78]. In each case the fit was checked by computing the normalized log-likelihood function. First the data point zero or close to zero was removed and then the normalized log-likelihood function computed for the remaining points.The experimental results are plotted in blue and the fits in red. The experimental results in Figs. 3.7 and 3.8 are from Sreenivasan and Dhruva [69] for the high Reynolds number atmospheric turbulence. The numbers plotted are from Table 2 in [19] where both experimental and simulation results are compared. We plotted the numbers from the latter simulation (1024^3) in Table 2, in [19]. We thank all of these researchers for the permission to use their results to compare with the theoretically computed PDFs. The NIG distribution was used by Barndorff-Nielsen et al. [8] to obtain fits to the PDFs for three different experimental data sets.

3.6 Description of Simulations and Experiments

First we described the simulations in the Ph.D. thesis of Michael Wilczek following [76]. The DNS data was produced by a standard pseudospectral code with periodic boundary conditions at a Taylor-based Reynolds number of 112. The simulations were run in a statistically stationary state with a large-scale forcing that preserves the kinetic energy of the flow and delivers approximately homogeneous isotropic turbulence. For more details we refer the reader to Michael Wilczek's Ph.D. thesis [76, 77].

The experiment by Xu et al. is described in their paper [78]: The turbulence is generated in a closed cylindrical chamber containing roughly 0.1 m^3 of water using counterrotating disks (French washing machine). The flow was seeded with transparent polystyrene microspheres with a diameter of 25 μm (smaller than or comparable to the smallest turbulent length scale) and a density 1.06 times that of water. These particles have previously been shown to act as passive tracers in this flow. The microspheres were illuminated by two pulsed Nd:YAG lasers, and their motion was recorded in three dimensions by three high-speed cameras at rates of up to 27,000 frames per second so that the smallest turbulent time scales were well resolved. The trajectories of individual tracer particles were reconstructed using particle tracking algorithms. Once the raw particle tracks were obtained, Lagrangian velocities were obtained by convolution with a Gaussian smoothing and differentiating kernel.

The smoothing operation works as a filter to suppress the measurement noise while the differentiation operation gives the derivative of the filtered signal.

The data from [69] consists of a series of measurements in atmospheric turbulence at Talylor microscale Reynolds number $\sim \sqrt{15R}$ ranging between 10,000 and 20,000. The Taylor frozen hypothesis is used, but it was verified by comparison with true spatial data obtained from two probes separated by a known streamwise distance; see [69]. The parameter values are listed in Table 3.1; see [19].

Table 3.1 Some relevant parameters for the atmospheric data

U	u'	ε	η	λ	R_λ
$7.6\,\mathrm{ms}^{-1}$	$1.36\,\mathrm{ms}^{-1}$	$0.032\,\mathrm{m}^2\,\mathrm{s}^{-3}$	0.57 mm	11.4 mm	10,340

Here, U is the mean speed, u' is the root-mean-square velocity, ε is the mean rate of energy dissipation, η and λ are the Kolmogorov and Taylor microscales, respectively, and $R_\lambda = u\lambda/\nu$, ν being the kinematic viscosity of air at the measurement temperature

Hotwire measurements were made in the atmospheric surface layer at a height of 35 m above the ground using a standard meteorological tower at Brookhaven National Laboratory. The tower itself presented very little obstacle to the wind because of its low solidity. The data set analyzed here is part of a more comprehensive batch of data obtained at the tower. The hotwire, 0.7 mm in length and 0.5 μm in diameter, was placed facing the wind, about two meters away from the tower. (For monitoring the wind direction, the tower was equipped with a vane anemometer placed two meters away from the measurement station.) The calibration was performed in situ using a TSI calibrator and checked later in a wind tunnel. The signals were low pass filtered at 5 kHz and sampled at 10 kHz. The anemometer and signal conditioners were placed nearby at the height of measurement, and the conditioned signal was transmitted to the ground and digitized using a 12-bit A/D converter. Typical data records contained between 10 and 40 million samples, during which time the wind direction and its mean speed were deemed acceptably constant. More details are given in Dhruva [22], but the essential features for this particular set of data are listed in Table 3.1. The wind conditions were somewhat unstable.

3.7 The Invariant Measure of the Stochastic Vorticity Equation

We first derive the stochastic vorticity equation from the stochastic Navier–Stokes equation (1.65),

$$
\begin{aligned}
u_t + u\cdot\nabla u &= \nu\Delta u - \nabla p + \sum_{k\neq 0} c_k^{\frac{1}{2}}\,\mathrm{d}b_t^k e_k(x) \\
&\quad + \sum_{k\neq 0} \mathrm{d}_k|k|^{1/3}\mathrm{d}t\,e_k(x) + u\sum_{k\neq 0}^{m}\int_{\mathbb{R}} h_k\bar{N}^k(\mathrm{d}t,\mathrm{d}z), \qquad (3.23)\\
u(x,0) &= u_0(x),
\end{aligned}
$$

with the incompressibility conditions

$$\nabla \cdot u = 0, \tag{3.24}$$

where $u(x),\ x \in \mathbb{R}$, is the velocity of the fluid and ν is the kinematic viscosity. Let $\omega = \nabla \times u$. Taking the curl of the Navier–Stokes equation (3.23) and using the vector identity

$$\nabla \times (u \cdot \nabla u) = u \cdot \nabla \omega - \omega \cdot \nabla u + (\nabla \cdot u)\omega = u \cdot \nabla \omega - \omega \cdot \nabla u,$$

because of the incompressibility condition (3.24), we get the vorticity equation

$$\begin{aligned} \omega_t + u \cdot \nabla \omega &= \nu \Delta \omega + \omega \cdot \nabla u + 2\pi \mathrm{i} \sum_{k \neq 0} k \times c_k^{\frac{1}{2}} \mathrm{d}b_t^k e_k(x) \\ &\quad + 2\pi \mathrm{i} \sum_{k \neq 0} k \times \mathrm{d}_k |k|^{1/3} \mathrm{d}t e_k(x) + \omega \sum_{k \neq 0}^{m} \int_{\mathbb{R}} h_k \bar{N}^k(\mathrm{d}t, \mathrm{d}z), \qquad (3.25) \\ \omega(x,0) &= \omega_0(x). \end{aligned}$$

This equation is linear in ω and we can solve it explicitly (in terms of u), using the Cameron-Martin formula (Girsanov's Theorem 2.2) and the Feynman–Kac formula, Lemma 2.3, as we did in Sects. 2.5 and 2.8. The solution is

$$\begin{aligned} \omega &= \mathrm{e}^{Kt} \mathrm{e}^{\int_0^t \mathrm{d}q} M_t \omega^0 + 2\pi \mathrm{i} \sum_{k \neq 0} k \times c_k^{1/2} \int_0^t \mathrm{e}^{K(t-s)} \mathrm{e}^{\int_s^t \mathrm{d}q} \mathrm{e}^{\int_s^t \nabla u \mathrm{d}r} M_{t-s} \mathrm{d}b_s^k e_k(x) \\ &+ 2\pi \mathrm{i} \sum_{k \neq 0} k \times \mathrm{d}_k \int_0^t \mathrm{e}^{K(t-s)} \mathrm{e}^{\int_s^t \mathrm{d}q} \mathrm{e}^{\int_s^t \nabla u \mathrm{d}r} M_{t-s} |k|^{1/3} \mathrm{d}s\, e_k(x), \end{aligned} \tag{3.26}$$

where K is the heat operator

$$K = \nu \Delta,$$

and

$$\begin{aligned} 3\int_s^t \mathrm{d}q &= \sum_{k \neq 0}^{M} \left\{ \int_0^t \int_{\mathbb{R}} \ln(1+h_k) \bar{N}^k(\mathrm{d}s, \mathrm{d}z) + \int_0^t \int_{\mathbb{R}} (\ln(1+h_k) - h_k) m_k(\mathrm{d}s, \mathrm{d}z) \right\} \\ &= \ln(|k|^{2/3} (2/3)^{N_t^k}), \end{aligned}$$

by Ito's formula and a computation similar to the one that produces the geometric Lévy process in Sect. 2.3, $m_k(\mathrm{d}t, \mathrm{d}z)$ being the kth Lévy measure; see Sect. 2.5. As in Sects. 2.5 and 2.8, M_t is the martingale

$$M_t = \exp\left\{ -\int_0^t u(B_s, s) \cdot \mathrm{d}B_s - \frac{1}{2} \int_0^t |u(B_s, s)|^2 \mathrm{d}s \right\}. \tag{3.27}$$

It is the Radon–Nikodym derivative of the measure, see Theorem 2.2, of the associated Ito process

$$\mathrm{d}x_t = -u \mathrm{d}t + \sqrt{2\nu} \mathrm{d}B_t,$$

u being the fluid velocity and $B_t = (B_t^1, B_t^2, B_t^3)$ a three-dimensional vector of auxiliary Brownian motions, with respect to the usual Brownian measure.

One can check that the vorticity does not depend on the velocity at any particular point (it only depends on the gradients of the velocity components). Similarly, we have used in the computation of (3.26) that the velocity at (x,t) is independent of the vorticity at the same point. The velocity only depends on the whole vorticity field through the Biot–Savart law

$$u(x,t) = -\frac{1}{4\pi}\int_{\mathbb{R}^3} \frac{(x-y)\times\omega(y,t)}{|x-y|^3}\mathrm{d}y, \tag{3.28}$$

where we have used the periodicity condition to extend the vorticity field to the whole of $\mathbb{R}^3$. The independence of $u(x,t)$ of $\omega(x,t)$ is seen by setting $\omega(x,t)=0$, since $\{x\}$ is a set of measure zero the integral in (3.28) is unchanged. The formula (3.26) also makes it clear that ω is an infinite-dimensional Ito–Lévy process; see [56]. Then by the Biot–Savart law u is also an infinite-dimensional Ito–Lévy process. This justifies the statements made in Sect. 2.5.

Now we define the variance

$$Q_t = \int_0^t P_s \mathrm{e}^{K(s)} C \mathrm{e}^{K^*(s)} P_s^* \mathrm{d}s \tag{3.29}$$

and drift

$$E_t = \int_0^t \mathrm{e}^{K(s)} P_s \bar{D} \mathrm{d}s \tag{3.30}$$

operators, where

$$P_t = \mathrm{e}^{\int_0^t \mathrm{d}q} \mathrm{e}^{\int_0^t \nabla u \mathrm{d}s} M_t = \prod_k^m (|k|^{2/3}(2/3)^{N_t^k})^{\frac{1}{3}} \mathrm{e}^{\int_0^t \nabla u \mathrm{d}s} M_t$$

and K is the heat operator. Then the Kolmogorov–Hopf equation for the vorticity can be written

$$\frac{\partial\phi}{\partial t} = \frac{1}{2}\mathrm{tr}[P_t C P_t^* \Delta\phi] + \mathrm{tr}[P_t \bar{D}\nabla\phi] + \langle K(\omega)P_t, \nabla\phi\rangle, \tag{3.31}$$

where $\bar{D} = (|k|^{1/3} D_k)$ and $\phi(\omega)$ is a bounded function of ω. The operators C and D are as in Sect. 3.1. It is also the Kolmogorov backward equation of the Ito process (3.26). The solution of (3.31) can be written in the form

$$\begin{aligned} R_t\phi(\omega) &= \int_H \phi(y) \mathcal{N}_{(\mathrm{e}^{Kt}P_t\omega + E_t I, Q_t)} * \mathbb{P}_{P_t}(\mathrm{d}y) \\ &= \int_H \phi(\mathrm{e}^{Kt}P_t\omega + E_t I + y)\mathcal{N}_{(0,Q_t)} * \mathbb{P}_{P_t}(\mathrm{d}y), \end{aligned}$$

where $\mathbb{P}_{P_t}$ is the Poisson law of P_t; see [56]. Here $|x| = \langle x,x\rangle^{1/2}$ where $\langle\cdot,\cdot\rangle$ is the inner product on H and $\omega = \omega_0$. $\mathcal{N}_{m,Q_t}$ denotes the normal distribution on H with mean m and variance Q_t, $I = \sum e_k$, and $E_t I \in H$.

3.7.1 The Invariant Measure of Turbulent Vorticity

We can now write a formula for the invariant measure of vorticity turbulence.

Theorem 3.3. *The invariant measure of the stochastic vorticity Navier–Stokes equation (3.25) on $H = L^2(\mathbb{T}^3)$ has the form*

$$\mu(\mathrm{d}x) = \mathrm{e}^{\langle Q^{-1/2}EI,\, Q^{-1/2}x\rangle - \frac{1}{2}|Q^{-1/2}EI|^2}\mathscr{N}_{(0,Q)}(\mathrm{d}x)\sum_k \delta_{k,l}\prod_{j\neq l}^{m}\delta_{N_t^j}\sum_{j=0}^{\infty} p_{m_l}^j \delta_{(N_t^l - j)}, \tag{3.32}$$

where $Q = Q_\infty$, $E = E_\infty$, $m_k = \ln|k|^{2/3}$ is the mean of the log-Poisson processes (2.26) and $p_{m_k}^j = \frac{(m_k)^j \mathrm{e}^{-m_k}}{j!}$ is the probability of $N_\infty^k = N_k$ having exactly j jumps, and $\delta_{k,l}$ is the Kronecker delta function.

Suppose that the operator Q is trace-class, $E(Q^{1/2}H) \subset Q^{1/2}(H)$, and that $\mathrm{e}^{Kt}P_t(H) \subset Q_t^{1/2}(H)$, $t > 0$, then, with u given, the invariant measure μ is unique, ergodic, and strongly mixing. We know that the above invariant measure is unique for the strong swirl [17] and strong rotation [3, 4], but it depends on u, and its uniqueness for general turbulent flows depends on the uniqueness of u.

The proof of Theorem 3.3 uses the above machinery and is analogous to the proof of Theorem 3.1 and Theorem 8.20 in [56].

The problem is that the vorticity may not be continuous although the velocity is. This is the reason why we use the Hilbert space $L^2(\mathbb{T}^3)$ in Theorem 3.3. In fact, we expect the vorticity to lack $1/3$ of a derivative see Corollary 4.2 below. This means that care must be taken when moments of the vorticity are computed and one may have to normalize the moments in order to get a finite answer. Nevertheless with proper normalization we can still project onto well-defined PDFs as in Sect. 3.3. The reason is that $\hat{c} = \lim_{k\to\infty}(Q^{-1/2}E)_k = |k \times \mathrm{d}_k||k|^{1/3}/|k \times c_k||k|^{1/3} = \bar{c}$, or the effect of the curl vanishes in the normalization, see Sect. 3.3. Therefore we still get the same stationary equation (3.16) for the PDF. Consequently, the four parameters in NIG in Sect. 3.4 are also the PDFs for the turbulent vorticity and its (normalized) moments.

Chapter 4
Existence Theory of Swirling Flow

4.1 Leray's Theory

We will consider the stochastic Navier–Stokes equation for the swirling flow (1.23), see Sect. 1.4, in the next three sections. Similar results hold for the stochastic Navier–Stokes equation (1.65) describing fully developed turbulence. However, to emphasize that (1.23) and (1.65) are not the same equations we will set the coefficients c_k to $c_k = h_k$ in (1.23) below. The h_ks can then be large but decay with increasing k. In this section we will first explain the probabilistic setting and prove some a priori estimates.

We let $(\Omega, \mathscr{F}, \mathbb{P})$, Ω is a set (of events) and $\mathscr{F}$ a σ-algebra on Ω, denote a probability space with $\mathbb{P}$ the probability measure of Brownian motion and $\mathscr{F}_t$ a filtration generated by all the Brownian motions b_t^k on $[t,\infty)$. If $f : \Omega \to H$ is a random variable, mapping Ω into a Hilbert space H, for example, $H = L^2(\mathbb{T}^3)$, then $L^2(\Omega, \mathscr{F}, \mathbb{P}; H)$ is a Hilbert space with norm:

$$\|f\|^2_{L^2(\Omega,\mathscr{F},\mathbb{P};H)} = E(|f(\omega)|_2^2) = \int_\Omega |f(\omega)|_2^2 \mathbb{P}(\mathrm{d}\omega) = \int_H |x|^2 f_\# \mathbb{P}(\mathrm{d}x),$$

where E denotes the expectation with respect to $\mathbb{P}$ and $f_\#\mathbb{P}$ denotes the pull back of the measure $\mathbb{P}$ to H. A stochastic process f_t in $\mathscr{L}^2 = L^2([0,T]; L^2(\Omega, \mathscr{F}, \mathbb{P}; H))$ has the norm

$$\|f_t\|^2_{\mathscr{L}^2} = \int_0^T E(|f(t,\omega)|_2^2)\mathrm{d}t$$

and f_t has the following properties; see [51].

Definition 4.1.

1. $f(t,\omega) : \mathbb{R}^+ \times \Omega \to \mathbb{R}$ is measurable with respect to $\mathscr{B} \times \mathscr{F}$ where $\mathscr{B}$ is the σ-algebra of the Borel sets on $[0,\infty)$, $\omega \in \Omega$.
2. $f(t,\omega)$ is adapted to the filtration $\mathscr{F}_t$.

B. Birnir, *The Kolmogorov-Obukhov Theory of Turbulence: A Mathematical Theory of Turbulence*, SpringerBriefs in Mathematics, DOI 10.1007/978-1-4614-6262-0_4,

3.

$$E\left(\int_0^T f^2(t,\omega)\mathrm{d}t\right) < \infty.$$

We are mostly interested in the Hilbert spaces $H = H^m(\mathbb{T}^3) = W^{(m,2)}$ that are the Sobolev spaces based on L^2 with the Sobolev norm

$$\|u\|_m^2 = |(1-\Delta^2)^{m/2}u|_2^2.$$

The corresponding norm on $\mathscr{L}_m^2 = L^2([0,T];L^2(\Omega,\mathscr{F},\mathbb{P};H^m(\mathbb{T}^3)))$ is

$$\|u\|_{\mathscr{L}_m^2} = \left[\int_0^T E(\|u\|_m^2)\mathrm{d}t\right]^{1/2}$$

more information about Sobolev spaces can be found in [1]. We will abuse notation slightly in this section by writing u instead of U; see Sect. 1.4. This is done for future reference and an easier comparison with Leray's classical estimates.

Let $\langle\cdot,\cdot\rangle$ denote the inner product on $L^2(\mathbb{T}^3)$. The following a priori estimates provide the foundation of the probabilistic version of Leray's theory.

Lemma 4.1. *The L^2 norms $|u|_2(\omega,t)$ and $|\nabla u|_2(\omega,t)$ satisfy the identity*

$$\mathrm{d}|u|_2^2 + 2\nu|\nabla u|_2^2\mathrm{d}t = 2\sum_{k\neq 0}\langle u, h_k^{1/2}e_k\rangle \mathrm{d}b_t^k + \sum_{k\neq 0} h_k\mathrm{d}t \tag{4.1}$$

and the bounds

$$|u|_2^2(\omega,t) \leq |u|_2^2(0)\mathrm{e}^{-2\nu\lambda_1 t} + 2\sum_{k\neq 0}\int_0^t \mathrm{e}^{-2\nu\lambda_1(t-s)}\langle u, h_k^{1/2}e_k\rangle \mathrm{d}b_s^k \tag{4.2}$$
$$+\frac{1-\mathrm{e}^{-2\nu\lambda_1 t}}{2\nu\lambda_1}\sum_{k\neq 0} h_k,$$

$$\int_0^t |\nabla u|_2^2(\omega,s)\mathrm{d}s \leq \frac{1}{2\nu}(|u|_2^2(0)-|\mathbf{U}|^2) + \frac{1}{\nu}\sum_{k\neq 0}\int_0^t \langle u, h_k^{1/2}e_k\rangle \mathrm{d}b_s^k + \frac{t}{2\nu}\sum_{k\neq 0} h_k, \tag{4.3}$$

where λ_1 is the smallest eigenvalue of $-\Delta$ with vanishing boundary conditions on the box $[0,1]^3$ and $h_k = |h_k^{1/2}|^2$. **U** *is the velocity vector from Sect. 1.4. The expectations of these norms are also bounded:*

$$E(|u|_2^2)(t) \leq E(|u|_2^2(0))\mathrm{e}^{-2\nu\lambda_1 t} + \frac{1-\mathrm{e}^{-2\nu\lambda_1 t}}{2\nu\lambda_1}\sum_{k\neq 0} h_k, \tag{4.4}$$

$$E\left(\int_0^t |\nabla u|_2^2(s)\mathrm{d}s\right) \leq \frac{1}{2\nu}[E(|u|_2^2(0)) - |\mathbf{U}|^2] + \frac{t}{2\nu}\sum_{k\neq 0} h_k. \tag{4.5}$$

Proof. The identity (4.1) follows from Leray's theory and Ito's lemma. We apply Ito's lemma to the L^2 norm of u squared:

$$d\int_{\mathbb{T}^3}|u|^2 dx = 2\int_{\mathbb{T}^3}\frac{\partial u}{\partial t}\cdot u dx dt + 2\sum_{k\neq 0}\int_{\mathbb{T}^3} u\cdot h_k^{1/2}e_k dx db_t^k + \sum_{k\neq 0} h_k\int_{\mathbb{T}^3} dx dt, \quad (4.6)$$

where $k\in\mathbb{Z}^3$ and $h_k^{1/2}\in\mathbb{R}^3$. Now by use of the Navier–Stokes equation (1.21)

$$\begin{aligned} d|u|_2^2 &= 2\int_{\mathbb{T}^3}\nu\Delta u\cdot u + (-u\cdot\nabla u + \nabla\Delta^{-1}(\text{trace}(\nabla u)^2)\cdot u dx dt \\ &\quad +2\sum_{k\neq 0}\int_{\mathbb{T}^3} u\cdot h_k^{1/2}e_k dx db_t^k + \sum_{k\neq 0} h_k dt \\ &= -2\nu|\nabla u|_2^2 dt + 2\sum_{k\neq 0}\int_{\mathbb{T}^3} u\cdot h_k^{1/2}e_k dx db_t^k + \sum_{k\neq 0} h_k dt \end{aligned}$$

since the divergent-free vector u is orthogonal both to the gradient $\nabla\Delta^{-1}(\text{trace}(\nabla u)^2)$ and $u\cdot\nabla u$ by the divergence theorem. Notice that the inner product (average) of u and the stirring force f in (1.21) vanish, $\langle u, f\rangle = \bar{u}\cdot f = 0$, so f can be omitted in the computation. The first term in the last expression is obtained by integration by parts. This is the identity (4.1). The inequality (4.2) is obtained by applying Poincaré's inequality

$$\lambda_1|u|_2^2 \leq |\nabla u|_2^2, \quad (4.7)$$

where λ_1 is the smallest eigenvalue of $-\Delta$ with vanishing boundary conditions on the cube $[0,1]^3$.[1] By Poincaré's inequality

$$\begin{aligned} d|u|_2^2 + 2\nu\lambda_1|u|_2^2 dt &\leq d|u|_2^2 + 2\nu|\nabla u|_2^2 dt \\ &= 2\sum_{k\neq 0}\langle u, h_k^{1/2}e_k\rangle db_t^k + \sum_{k\neq 0} h_k dt. \end{aligned}$$

Solving the inequality gives (4.2). Equation (4.3) is obtained by integrating (4.1)

$$|u|_2^2(t) + 2\nu\int_0^t|\nabla u|_2^2(s)ds = |u|_2^2(0) + 2\sum_{k\neq 0}\int_0^t\langle u, h_k^{1/2}e_k\rangle db_s^k + t\sum_{k\neq 0} h_k$$

and dropping $|u-\mathbf{U}|_2^2(t) > 0$, by use of (1.37).

Finally we take the expectations of (4.2) and (4.3) to obtain, respectively, (4.4) and (4.5), using that the function $\langle u, h_k^{1/2}e_k\rangle(\omega,t)$ is adapted to the filtration $\mathscr{F}_t$.

The following amplification of Leray's a priori estimates will play an important role in the a priori estimates of the solution of the stochastic Navier–Stokes equation below.

[1] We should subtract the mean from u in Poincaré's inequality because of the periodic boundary conditions, but the mean just washes out in the estimates.

Lemma 4.2. *Let $u_{\frac{1}{2B}} = u(x,t+\frac{1}{2B})$ denote the translation of u in time by the number $\frac{1}{2B}$. Then the L^2 norms of the differences $|u-u_{\frac{1}{2B}}|_2(\omega,t)$ and $|\nabla u-\nabla u_{\frac{1}{2B}}|_2(\omega,t)$ satisfy the identity*

$$d|u-u_{\frac{1}{2B}}|_2^2+2\nu|\nabla u-\nabla u_{\frac{1}{2B}}|_2^2 dt = 2\sum_{k\neq 0}\langle u-u_{\frac{1}{2B}}, h_k^{1/2}e_k\rangle d(b_t^k - b_{t+\frac{1}{2B}}^k) \tag{4.8}$$

and the bounds

$$|u-u_{\frac{1}{2B}}|_2^2(\omega,t) \leq |u-u_{\frac{1}{2B}}|_2^2(0)e^{-2\nu\lambda_1 t} + 2\sum_{k\neq 0}\int_0^t e^{-2\nu\lambda_1(t-s)}\langle u-u_{\frac{1}{2B}}, h_k^{1/2}e_k\rangle d(b_s^k - b_{t+\frac{1}{2B}}^k) \tag{4.9}$$

$$\int_0^t |\nabla u-\nabla u_{\frac{1}{2B}}|_2^2(\omega,s)ds \leq \frac{1}{2\nu}|u-u_{\frac{1}{2B}}|_2^2(0) + \frac{1}{\nu}\sum_{k\neq 0}\int_0^t \langle u-u_{\frac{1}{2B}}, h_k^{1/2}e_k\rangle d(b_s^k - b_{t+\frac{1}{2B}}^k), \tag{4.10}$$

where λ_1 is the smallest eigenvalue of $-\Delta$ with vanishing boundary conditions on the box $[0,1]^3$ and $h_k = |h_k^{1/2}|^2$. The expectations of these norms are also bounded

$$E(|u-\nabla u_{\frac{1}{2B}}|_2^2)(t) \leq E(|u-\nabla u_{\frac{1}{2B}}|_2^2(0))e^{-2\nu\lambda_1 t} \tag{4.11}$$

$$E\left(\int_0^t |\nabla u-\nabla u_{\frac{1}{2B}}|_2^2(s)ds\right) \leq \frac{1}{2\nu}E(|u-\nabla u_{\frac{1}{2B}}|_2^2(0)) \tag{4.12}$$

by the expectations of the initial data of the differences.

The proof of this lemma is analogous to the proof of Lemma 4.1 and can be found in [17].

Remark 4.1. Notice that in the notation of Sect. 1.4 $|U-U_{\frac{1}{2B}}|_2^2 = |u-u_{\frac{1}{2B}}|_2^2$ because the constant velocity **U** cancels out.

4.2 The A Priori Estimate of the Turbulent Solutions

The mechanism of the turbulence production are fast oscillations driving large turbulent noise that was initially seeded by small white noise, as explained in the previous section. These fast oscillations are generated by the fast constant flow $U = U_1$, where we have dropped the subscript 1, and the flow is rotating with amplitude A and angular velocity Ω. The frequency of these oscillations increases with U and $A\Omega$. The bigger U and $A\Omega$ are the more efficient this turbulence production mechanism becomes.

In this section we will establish an a priori estimate on the norm of the turbulent solution that allows us to extend the local existence and uniqueness to the whole real-time axis. Thus the a priori estimates suffice to give global existence and uniqueness. We recall the oscillatory kernel (1.34) from Sect. 1.4:

$$\sum_{k\neq 0} h_k^{1/2} \int_0^t \mathrm{e}^{-(4\pi^2|k|^2+2\pi \mathrm{i} U_1 k_1)(t-s)-2\pi \mathrm{i} A(k_2,k_3)[\sin(\Omega t+\theta)-\sin(\Omega s+\theta)]} \mathrm{d}b_s^k e_k(x). \quad (4.13)$$

The imaginary part of the argument of the exponential creates oscillations and as U_1 and $A\Omega$ become larger these oscillations become faster. We take advantage of this mechanism to produce the a priori estimates.

Next lemma plays a key role in the proof of the useful estimate of the turbulent solution. It is a version of the Riemann–Lebesgue lemma which captures the averaging effect (mixing) of the oscillations.

Lemma 4.3. *Let the Fourier transform in time be*

$$\tilde{w} = \int_0^T w(s)\mathrm{e}^{-2\pi \mathrm{i}(k_1 U + A(k_2,k_3)\Omega)s} \mathrm{d}s,$$

where $A(k_2,k_3) = A\sqrt{k_2^2+k_3^2}$ *and* $w = w(k,t)$, $k = (k_1,k_2,k_3)$, *is a vector with three components. If* T *is an even integer multiple of* $\frac{1}{k_1 U + A(k_2,k_3)\Omega}$, *then*

$$\tilde{w} = \widetilde{\#w}, \quad (4.14)$$

where

$$\#w = \frac{1}{2}\left[w(s) - w\left(s + \frac{1}{2[k_1 U + A(k_2,k_3)\Omega]}\right)\right] = \frac{1}{2}\int_{s+\frac{1}{2|k_1 U + A(k_2,k_3)\Omega|}}^{s} \frac{\partial w}{\partial r} \mathrm{d}r \quad (4.15)$$

and $\#w$ *satisfies the estimate*

$$|\#w| \leq \frac{1}{4|k_1 U + A(k_2,k_3)\Omega|} \operatorname{ess\,sup}_{[s,s+\frac{1}{2(k_1 U_1 + A(k_2,k_3)\Omega)}]} \left|\frac{\partial w}{\partial s}\right|. \quad (4.16)$$

Proof. The proof is similar to the proof of the Riemann–Lebesgue lemma for the Fourier transform in time, let $B(k) = k_1 U + A(k_2,k_3)\Omega$:

$$\begin{aligned}
\tilde{w}(k) &= \int_0^T w(s)\mathrm{e}^{-2\pi \mathrm{i} B s} \mathrm{d}s \\
&= -\int_0^T w(s)\mathrm{e}^{-2\pi \mathrm{i} B (s-\frac{1}{2B})} \mathrm{d}s \\
&= -\int_0^T w\left(s+\frac{1}{2B}\right)\mathrm{e}^{-2\pi \mathrm{i} B s} \mathrm{d}s,
\end{aligned}$$

where we have used in the last step that w is a periodic function on the interval $[0,T]$. Taking the average of the first and the last expression we get

$$\tilde{w} = \frac{1}{2}\int_0^T \left(w(s) - w\left(s+\frac{1}{2B}\right)\right) \mathrm{e}^{-2\pi \mathrm{i} Bs} \mathrm{d}s = \widetilde{\#w}.$$

Now

$$\begin{aligned}
|\#w| &= \frac{1}{2}\left|\left(w(s) - w\left(s+\frac{1}{2B}\right)\right)\right| \\
&\leq \frac{1}{2}\int_s^{s+\frac{1}{2B}} \left|\frac{\partial w}{\partial r}\right| \mathrm{d}r \\
&\leq \frac{1}{4|B|} \text{ess sup}_{[s,s+\frac{1}{2B}]} \left|\frac{\partial w}{\partial s}\right|
\end{aligned}$$

by the mean-value theorem.

Corollary 4.1. *If T is not an even integer multiple of $\frac{1}{B(k)} = \frac{1}{k_1 U + A(k_2,k_3)\Omega}$, then*

$$\tilde{w} = \widetilde{\#w} - \frac{1}{2}\int_{-\frac{1}{2B}}^0 w\left(s+\frac{1}{2B}\right) \mathrm{e}^{-2\pi \mathrm{i} Bs} \mathrm{d}s + \frac{1}{2}\int_{T-\frac{1}{2B}}^T w\left(s+\frac{1}{2B}\right) \mathrm{e}^{-2\pi \mathrm{i} Bs} \mathrm{d}s, \quad (4.17)$$

where $\tilde{w}$ satisfies the estimate

$$|\tilde{w}| \leq |\widetilde{\#w}| + \frac{1}{|B|} \text{ess sup}_{[-\frac{1}{2B},0] \cap [T-\frac{1}{2B},T]} \left|w\left(s+\frac{1}{2B}\right)\right|. \quad (4.18)$$

Proof. The proof is the same as of the lemma except for the step

$$\begin{aligned}
\tilde{w}(k) &= \int_0^T w(s) \mathrm{e}^{-2\pi \mathrm{i} Bs} \mathrm{d}s = -\int_0^T w(s) \mathrm{e}^{-2\pi \mathrm{i} B(s-\frac{1}{2B})} \mathrm{d}s \\
&= -\int_0^T w(s+\frac{1}{2B}) \mathrm{e}^{-2\pi \mathrm{i} Bs} \mathrm{d}s - \int_{-\frac{1}{2B}}^0 w\left(s+\frac{1}{2B}\right) \mathrm{e}^{-2\pi \mathrm{i} Bs} \mathrm{d}s \\
&\quad + \int_{T-\frac{1}{2B}}^T w\left(s+\frac{1}{2B}\right) \mathrm{e}^{-2\pi \mathrm{i} Bs} \mathrm{d}s.
\end{aligned}$$

The lemma allows us to estimate the Fourier transform (in t) of w in terms of the time derivative of w, with a gain of $(k_1 U + A(k_2,k_3)\Omega)^{-1}$. Below we will use it in an estimate showing that the limit of $\#w$ is zero when $|B(k)| = |(k_1 U + A(k_2,k_3)\Omega)| \to \infty$.

Lemma 4.4. *The integral*

$$\int_0^t (2\pi|k|)^p \mathrm{e}^{-(4\pi^2 \nu |k|^2 + 2\pi \mathrm{i}[B(k)(t-s)+g])} \mathrm{d}s,$$

where $B(k) = k_1 U + A(k_2, k_3)\Omega$, is bounded by

$$(2\pi)^p \int_0^t |k|^p e^{-4\pi^2 \nu |k|^2 (t-s)} ds \leq C\, t^{1-\frac{p}{2}} \tag{4.19}$$

for $0 \leq p < 2$, where C is a constant. In particular,

$$\int_{t-\delta}^t (2\pi|k|)^p e^{-(4\pi^2\nu|k|^2 + 2\pi i[B(k)(t-s)+g])} ds \leq C\, \delta^{1-\frac{p}{2}}. \tag{4.20}$$

Proof. We estimate the integral

$$\begin{aligned}\int_0^t |k|^p e^{-4\pi^2\nu|k|^2(t-s)} ds &= \int_0^t |k|^p e^{-4\pi^2\nu|k|^2 r} dr \\ &\leq \left(\frac{p}{4\pi^2}\right)^{\frac{p}{2}} e^{-p} \int_0^t r^{-\frac{p}{2}} dr = Ct^{1-\frac{p}{2}},\end{aligned}$$

where

$$k = \frac{1}{2\pi}\sqrt{\frac{p}{r}}$$

is the value of k where the integrand achieves its maximum.

The rotation can resonate with the uniform (linear) flow due to the nonlinearities in the Navier–Stokes equation. The following lemma restricts the values of velocity coefficients so that no resonance occurs.

Lemma 4.5. *Suppose that for $k_1 < 0$ and $\frac{\sqrt{k_2^2+k_3^2}}{|k_1|} \neq 0$ or ∞, the constants U, A, and Ω satisfy the non-resonance condition*

$$\left| \frac{U}{A\Omega} + \frac{\sqrt{k_2^2 + k_3^2}}{k_1} \right| \geq \frac{C}{|k_1|^r}, \tag{4.21}$$

where C is a constant and $0 < r < 1$; then for all $k = (k_1, k_2, k_2) \neq 0$,

$$|Uk_1 + A\Omega\sqrt{k_2^2 + k_3^2}| \neq 0 \tag{4.22}$$

and

$$\lim_{|k|\to\infty} |Uk_1 + A\Omega\sqrt{k_2^2 + k_3^2}| = \infty. \tag{4.23}$$

Moreover,

$$|Uk_1 + A\Omega\sqrt{k_2^2 + k_3^2}| \geq B = \min(U, A\Omega, CA\Omega). \tag{4.24}$$

Proof. If $k_1 > 1$, then

$$\left|Uk_1 + A\Omega\sqrt{k_2^2+k_3^2}\right| = U|k_1| + A\Omega\sqrt{k_2^2+k_3^2} > 0$$

so (4.22) and (4.23) hold. If $k_1 < 0$, then by (4.21)

$$\left|Uk_1 + A\Omega\sqrt{k_2^2+k_3^2}\right| \geq C\,\Omega A|k_1|^{1-r} > 0$$

and

$$\lim_{|k|\to\infty}\left|Uk_1 + A\Omega\sqrt{k_2^2+k_3^2}\right| \geq C\,\Omega A \lim_{|k_1|\to\infty}|k_1|^{1-r} = \infty$$

if $|k_1| \to \infty$. If on the other hand $|k_1| < \infty$ when $|k| \to \infty$ then (4.23) also holds. When $k_1 = 0$, (4.22) and (4.23) are obvious and also if $k_2 = k_3 = 0$.

The lower bound (4.24) is read of

$$\left|Uk_1 + A\Omega\sqrt{k_2^2+k_3^2}\right|$$

when $k_1 \geq 1$. Then it is either U or $A\Omega$. When $k_1 = 0$ then it is $A\Omega$ and by (4.21), when $k_1 \leq -1$, it is greater than or equal $CA\Omega$.

The next question to ask is in which space do the turbulent solutions live? This was pointed out by Onsager in 1945 [53]. He pointed out that if the solutions satisfy the Kolmogorov scaling down to the smallest scales, they must be Hölder continuous function with Hölder exponent 1/3. In three dimensions this means that they live in the Sobolev space $H^{\frac{11}{6}+\varepsilon}$ based on $L^2(\mathbb{T}^3)$.

If $\frac{q}{p}$ is a rational number let $\frac{q}{p}^+$ denote any real number $s > \frac{q}{p}$.

Theorem 4.1. *Let the velocity $U = U_1$ of the mean flow and the product $A\Omega$ of the amplitude A and the frequency Ω of the rotation be sufficiently large, in the uniform rotating flow (1.19), with U, $A\Omega$ also satisfying the non-resonance conditions (4.21). Then the solution of the integral equation (1.32) is uniformly bounded in $\mathcal{L}^2_{\frac{11}{6}^+}$,*

$$\text{ess sup}_{t\in[0,\infty)} E(\|u\|^2_{\frac{11}{6}^+})(t) \leq \left(1 - C\left(\frac{1}{B^2} + \delta^{\frac{1}{6}^-}\right)\right)^{-1}\left[\sum_{k\neq 0}\frac{3(1+(2\pi|k|)^{\frac{11}{3}^+})}{8\pi^2\nu|k|^2}h_k + \frac{C'}{B}\right], \tag{4.25}$$

where $B = \min(|U|, A\Omega, CA\Omega)$ is large, δ small, and C and C' are constants.

Corollary 4.2 (Onsager's Observation). *The solutions of the integral equation (1.32) are Hölder continuous with exponent* $1/3$.

Remark 4.2. The estimate (4.25) provides the answer to the question we posed in Sect. 1.4 how fast the coefficients $h_k^{1/2}$ had to decay in Fourier space. They have to decay sufficiently fast for the expectation of the $H^{\frac{11}{6}^+} = W^{(\frac{11}{6}^+,2)}$ Sobolev norm of the initial function u_0, to be finite. This expectation appear on the right-hand side of (4.25). In other words the $\mathscr{L}^2_{\frac{11}{6}^+}$ norm of the initial function u_0 has to be finite.

The proof of the theorem involves long estimates and can be found in [17]. An outline of the proof is given in Appendix A.

We consider the integral equation

$$u(x,t) = \sum_{k\neq 0}\Bigg[h_k^{1/2}A_t^k - \int_0^t e^{-[4\pi^2\nu|k|^2+2\pi i B(k)](t-s)-2\pi i g(k,t,s)} \times\left(\widehat{u\cdot\nabla u}+\frac{ik}{2\pi|k|^2}(\widehat{\mathrm{tr}(\nabla u)^2})\right)(k,s)ds\Bigg]e_k(x),$$

where $B(k) = Uk_1 + A(k_2,k_3)\Omega$.

Lemma 4.6. *The initial condition $(u-u_{\frac{1}{2B}})(0)$ satisfies the estimate*

$$|u-u_{\frac{1}{2B}}|_2^2(0) \le 2\sum_{j\neq 0}|A^j_{\frac{1}{2B(k)}}|^2 + \frac{C}{|B(k)|^2}\text{ess sup}_{t\in[0,\frac{1}{2B}]}\|u\|^2_{\frac{11}{6}^+}. \tag{4.26}$$

Proof. We use the integral equation

$$\begin{aligned} u-u_{\frac{1}{2B}} = \sum_{k\neq 0}\Bigg[&h_k^{1/2}(A_t^k - A^k_{t+\frac{1}{2B}}) \\ &-\left(\int_0^t e^{-[4\pi^2\nu|k|^2+2\pi i B(k)](t-s)-2\pi i g(k,t,s)}\right) \\ &\times\left(\widehat{u\cdot\nabla u}+\frac{ik}{2\pi|k|^2}(\widehat{\mathrm{tr}(\nabla u)^2})\right)(k,s)ds \\ &-\int_0^{t+\frac{1}{2B}} e^{-[4\pi^2\nu|k|^2+2\pi i B(k)](t+\frac{1}{2B}-s)-2\pi i g(k,t+\frac{1}{2B},s)} \\ &\times\left(\widehat{u\cdot\nabla u}+\frac{ik}{2\pi|k|^2}(\widehat{\mathrm{tr}(\nabla u)^2})\right)(k,s)ds\Bigg]e_k(x), \end{aligned}$$

where $B(k) = Uk_1 + A(k_2,k_3)\Omega$. At $t = 0$,

$$|u-u_{\frac{1}{2B}}|^2(0) = |u_{\frac{1}{2B}}|^2(0) = 2\sum_{j\neq 0}h_j|A^j_{\frac{1}{2B}}|^2 + \frac{C}{|B(k)|^2}\text{ess sup}_{t\in[0,\frac{1}{2B}]}\|u\|^2_{\frac{11}{6}^+}$$

by the same estimates as above.

Lemma 4.7. *The identity (4.1) in Lemma 4.1 can be modified for $a > 0$*

$$d(e^{\nu at}|u|_2^2) + 2\nu e^{\nu at}|\nabla u|_2^2 dt = \nu a e^{\nu at}|u|_2^2 dt + 2e^{\nu at}\sum_{k\neq 0}\langle u, h_k^{1/2}e_k\rangle db_t^k + e^{\nu at}\sum_{k\neq 0} h_k dt \tag{4.27}$$

and produces the estimates

$$\begin{aligned}|u|_2^2(t) \leq |u|_2^2(0)\left(e^{-\nu at} + \frac{ae^{-2\nu\lambda_1 t}}{(a-2\lambda_1)}\right) &+ 2\sum_{k\neq 0}\int_0^t e^{-\nu a(t-s)}\langle u, h_k^{1/2}e_k\rangle db_s^k \\ +2\sum_{k\neq 0}\int_0^t e^{-\nu a(t-s)}\int_0^s e^{-2\nu\lambda_1(s-r)}\langle u, h_k^{1/2}e_k\rangle db_r^k ds &+ \frac{1}{\nu}\left(\frac{1}{a} + \frac{1}{2\lambda_1}\right)\sum_{k\neq 0} h_k\end{aligned} \tag{4.28}$$

and

$$\begin{aligned}\int_0^t e^{-\nu a(t-s)}|\nabla u|_2^2(s)ds \leq \frac{1}{2\nu}(|u|_2^2(0) - |\mathbf{U}|^2)&\left(e^{-\nu at} + \frac{ae^{-2\nu\lambda_1 t}}{(a-2\lambda_1)}\right) \\ &+\frac{1}{\nu}\sum_{k\neq 0}\int_0^t e^{-\nu a(t-s)}\langle u, h_k^{1/2}e_k\rangle db_s^k \\ &+\frac{1}{\nu}\sum_{k\neq 0}\int_0^t e^{-\nu a(t-s)}\int_0^s e^{-2\nu\lambda_1(s-r)}\langle u, h_k^{1/2}e_k\rangle db_r^k ds \\ &+\frac{1}{2\nu^2}\left(\frac{1}{a} + \frac{1}{2\lambda_1}\right)\sum_{k\neq 0} h_k,\end{aligned} \tag{4.29}$$

where λ_1 is the smallest eigenvalue of $-\Delta$ with vanishing boundary conditions on the box $[0,1]^3$ and $h_k = |h_k^{1/2}|^2$.

Proof. We multiply the identity (4.1) in Lemma 4.1 by $e^{\nu at}$ to get (4.27). Then integration gives the equality

$$\begin{aligned}|u|_2^2(t) + 2\nu\int_0^t e^{-\nu a(t-s)}|\nabla u|_2^2(s)ds = |u|_2^2(0)e^{-\nu at} &+ \nu a\int_0^t e^{-\nu a(t-s)}|u|_2^2(s)ds \\ &+2\sum_{k\neq 0}\int_0^t e^{-\nu a(t-s)}\langle u, h_k^{1/2}e_k\rangle db_s^k \\ &+\frac{(1-e^{-\nu a(t-s)})}{\nu a}\sum_{k\neq 0} h_k.\end{aligned}$$

Now substituting the estimate (4.2), from Lemma 4.1, for $|u|_2^2$ on the right-hand side gives the two inequalities (4.28) and (4.29) as in Lemma 4.1.

Lemma 4.8. *The functions H, K, and L in the proof of Theorem 4.1 satisfy the estimate*

$$E(H+K+L) \leq \frac{C}{|B(k)|^2} E(\text{ess sup}_{t\in[0,\frac{1}{2B}]} \|u\|^2_{\frac{11}{6}^+}) + \frac{C'}{B} \tag{4.30}$$

with $B = \min(U, A\Omega, CA\Omega)$.

The proof of the lemma involves long formulas for H, K, and L and can be found in [17].

Remark 4.3. Corollary 4.2 is the resolution of a famous question in turbulence, for the swirling flows: *Is turbulence always caused by the blow up of the velocity u?* The answer according to Theorem 4.1 is *no*; the solutions are not singular. However, they are not smooth either, contrary to the belief, stemming from Leray's theory [42], that if solutions are not singular then they are smooth. By Corollary 4.2 the solutions are Hölder continuous with exponent $1/3$ in three dimensions. This confirms an observation made by Onsager [54] in 1945. In particular the gradient ∇u and vorticity $\nabla \times u$ are not continuous in general as discussed in Sect. 3.7.

Remark 4.4. U and $A\Omega$ do not have to be made very large for the estimate (4.25) to be satisfied, because $B(k) \to \infty$ as $|k| \to \infty$. How big U and $A\Omega$ have to be for (4.25) to hold is probably best answered by a numerical simulation.

We can now prove that $\text{ess sup}_{t\in[0,\infty)} \|u(t)\|^2_{\frac{11}{6}^+}$ is bounded with probability close to one.

Lemma 4.9. *For all $\varepsilon > 0$ there exists an R such that*

$$\mathbb{P}(\text{ess sup}_{t\in[0,\infty)} \|u(t)\|^2_{\frac{11}{6}^+} < R) > 1 - \varepsilon. \tag{4.31}$$

Proof. By Chebyshev's inequality and the estimate (4.25) we get that

$$\mathbb{P}(\text{ess sup}_{t\in[0,\infty)} \|u(t)\|^2_{\frac{11}{6}^+} \geq R) < \frac{C}{R} < \varepsilon$$

for R sufficiently large.

4.3 Existence Theory of the Stochastic Navier–Stokes Equation

In this section we prove the existence of the turbulent solutions of the initial value problem (1.23). The following theorem states the existence of turbulent solutions in three dimensions. First we write the initial value problem (1.23) as the integral equation (4.32)

$$u(x,t) = u_0(x,t) - \int_0^t e^{K(t-s)} * [u \cdot \nabla u - \nabla \Delta^{-1} \mathrm{tr}(\nabla u)^2] \mathrm{d}s. \qquad (4.32)$$

Here e^{Kt} is the oscillatory heat kernel (1.33) and

$$u_0(x,t) = \sum_{k \neq 0} h_k^{1/2} A_t^k e_k(x)$$

the A_t^ks being the oscillatory Ornstein–Uhlenbeck-type processes from (1.34).

Theorem 4.2. *If the uniform flow U and product of the amplitude and frequency $A\Omega$, of the rotation, are sufficiently large, $B = \min(|U|, A\Omega, CA\Omega)$, δ is small and the non-resonance conditions (4.21) are satisfied, so that the a priori bound (4.25) holds, then the integral equation (4.32) has unique global solution $u(x,t)$ in the space $C([0,\infty); L^2(\Omega, \mathscr{F}, \mathbb{P}; H^{\frac{11}{6}^+}))$, u is adapted to the filtration generated by the stochastic process*

$$u_0(x,t) = \sum_{k \neq 0} h_k^{1/2} A_t^k e_k$$

and

$$E\left(\int_0^t \|u\|^2_{\frac{11}{6}^+} \mathrm{d}s\right) \leq \left(1 - C\left(\frac{1}{B^2} + \delta^{\frac{1}{6}^-}\right)\right)^{-1} \left[\sum_{k \neq 0} \frac{3(1 + (2\pi|k|)^{\frac{11}{3}^+})}{8\pi^2 \nu |k|^2} h_k + \frac{C'}{B}\right] t. \qquad (4.33)$$

This theorem is a standard application of the contraction mapping principle to prove global existence and uniqueness. Then the unique local solution is extended to the whole positive time axis by use of the a priori bound (4.25). A detailed proof can be found in [17].

We now add the initial condition $u(x,0) = u^0(x)$, with mean zero, to the integral equation (4.32).

Theorem 4.3. *If the uniform flow U and the product of the amplitude $A\Omega$ and frequency of the rotation, $B = \min(|U|, A\Omega, CA\Omega)$, are sufficiently large, δ small, and the non-resonance conditions (4.21) are satisfied, so that the a priori bound (4.25) holds, then the integral equation*

$$u(x,t) = e^{Kt} * u^0(x) + u_0(x,t) - \int_0^t e^{K(t-s)} * (u \cdot \nabla u - \nabla \Delta^{-1} (\nabla u)^2)\, \mathrm{d}s, \qquad (4.34)$$

where e^{Kt} is the oscillating kernel in (1.33), has unique global solution $u(x,t)$ in the space $C([0,\infty); L^2(\Omega, \mathscr{F}, \mathbb{P}; H^{\frac{11}{6}^+}))$, u is adapted to the filtration generated by the stochastic process

$$u_0(x,t) = \sum_{k \neq 0} h_k^{1/2} A_t^k e_k$$

and

$$E\left(\int_0^t \|u\|^2_{\frac{11}{6}^+} ds\right) \le \left(1 - C\left(\frac{1}{B^2} + \delta^{\frac{1}{6}^-}\right)\right)^{-1} \left[\sum_{k\neq 0} \frac{(1+(2\pi|k|)^{\frac{11}{3}^+})}{2\pi^2\nu|k|^2} h_k + \frac{C'}{B}\right] t. \tag{4.35}$$

The proof of the theorem is exactly the same as the proof of Theorem 4.2 once the a priori bound (4.25) is established. A proof can be found in [17].

Corollary 4.3. *For any initial data $u^0 \in \dot{L}^2(\mathbb{T}^3)$, the L^2 space with mean zero, and any $t_0 > 0$, there exists a mean flow U, an amplitude and angular velocity $A\Omega$, and δ small, such that (4.34) has a unique solution in $C([t_0,\infty); L^2(\Omega,\mathscr{F},\mathbb{P}; H^{\frac{11}{6}^+}))$.*

Proof. For $t > 0$, $e^{Kt} * u^0(x)$ is smooth. Now apply Theorem 4.3.

Next we prove a Gronwall estimate that can be use to prove local (in t) stability and irreducibility; see [17].

Lemma 4.10. *Let u be a solution of (4.32) with an initial function $u_0(x,t) = \sum_{k\neq 0} h_k^{1/2} A_t^k e_k$ and initial condition $u^0(x)$ and y a solution of*

$$y_t + \mathbf{U}\cdot\nabla y = \nu\Delta y - y\cdot\nabla y + \nabla\Delta^{-1}\mathrm{tr}(\nabla y)^2 + f \tag{4.36}$$

with initial condition $y^0(x)$, then

$$\begin{aligned}\|u-y\|^2_{\frac{11}{6}^+}(t) \le{}& [3\|u^0 - y^0\|^2_{\frac{11}{6}^+} + 3\|\sum_{k\neq 0} h_k^{1/2} A_t^k e_k - e^{Kt} * f\|^2_{\frac{11}{6}^+} \\ &+ \delta^2 C_1 \operatorname{ess\,sup}_{s\in[t-\delta,t]}(\|u\|^2_{\frac{11}{6}^+} + \|y\|^2_{\frac{11}{6}^+})] e^{C_2\int_0^{t-\delta}(1+\|u\|^2_{\frac{11}{6}^+} + \|y\|^2_{\frac{11}{6}^+})ds},\end{aligned} \tag{4.37}$$

where C_1 and C_2 are constants and δ can be made arbitrarily small. The A_t^ks are the oscillatory Ornstein–Uhlenbeck-type processes (1.35) and e^{Kt} is the oscillatory kernel in (1.33).

Proof. We subtract the integral equation for y from that of u:

$$\begin{aligned} u &= u^0 + \sum_{k\neq 0} h_k^{1/2} A_t^k e_k + e^{Kt} * (-u\cdot\nabla u + \nabla\Delta^{-1}\mathrm{tr}(\nabla u)^2), \\ y &= y^0 + e^{Kt} * f + e^{Kt} * (-y\cdot\nabla y + \nabla\Delta^{-1}\mathrm{tr}(\nabla y)^2). \end{aligned}$$

Thus

$$\begin{aligned}\|u-y\|^2_{\frac{11}{6}^+}(t) \le{}& [3\|u^0 - y^0\|^2_{\frac{11}{6}^+} + 3\|\sum_{k\neq 0} h_k^{1/2} A_t^k e_k - e^{Kt} * f\|^2_{\frac{11}{6}^+} \\ &+ 3\|e^{Kt} * (-w\nabla u - y\nabla w + \nabla\Delta^{-1}\mathrm{tr}\nabla\alpha\cdot\nabla w)\|^2_{\frac{11}{6}^+}],\end{aligned}$$

where $w = u - y$ and $\alpha = u + y$. Now the same estimates as in Theorem 4.1 give

$$\begin{aligned}\|u-y\|^2_{\frac{11}{6}+}(t) \le 3\|u^0-y^0\|^2_{\frac{11}{6}+} + 3\|\sum_{k\neq 0} h_k^{1/2} A_t^k e_k - \mathrm{e}^{Kt} * f\|^2_{\frac{11}{6}+} \\ +C_1\delta^2 \text{ess sup}_{s\in[t-\delta,t]}(\|u\|^2_{\frac{11}{6}+} + \|y\|^2_{\frac{11}{6}+}) \\ +C_2\int_0^{t-\delta}(1+\|u\|^2_{\frac{11}{6}+} + \|y\|^2_{\frac{11}{6}+})(\|u-y\|^2_{\frac{11}{6}+})\mathrm{d}s.\end{aligned}$$

Then Grönwall's inequality gives (4.37).

Appendix A
The Bound for a Swirling Flow

Theorem A.1. *Let the velocity* $U = U_1$ *of the mean flow and the product* $A\Omega$ *of the amplitude* A *and the frequency* Ω *of the rotation be sufficiently large, in the uniform rotating flow (1.19), with* U, $A\Omega$ *also satisfying the non-resonance conditions (4.21). Then the solution of the integral equation (1.32) is uniformly bounded in* $\mathscr{L}^2_{\frac{11}{6}+}$,

$$\text{ess sup}_{t\in[0,\infty)} E\left(\|u\|^2_{\frac{11}{6}+}\right)(t) \le \left(1 - C\left(\frac{1}{B^2} + \delta^{\frac{1}{6}^-}\right)\right)^{-1} \times \left[\sum_{k\neq 0} \frac{3(1+(2\pi|k|)^{\frac{11}{3}^+})}{8\pi^2 \nu |k|^2} h_k + \frac{C'}{B}\right], \tag{A.1}$$

where $B = \min(|U|, A\Omega, CA\Omega)$ *is large,* δ *small, and* C *and* C' *are constants.*

Corollary A.1 (Onsager's Observation). *The solutions of the integral equation (1.32) are Hölder continuous with exponent* $1/3$.

Outline of Proof: We write the integral equation (1.32) in the form

$$u(x,t) = \sum_{k\neq 0} \left[h_k^{1/2} A_t^k - \int_0^t e^{-(\{4\pi^2\nu|k|^2 + 2\pi i[k_1 U_1 + A(k_2,k_3)\Omega]\}(t-s) + 2\pi i g(k,t,s))} \times (\widehat{u\cdot\nabla u} - \nabla\Delta^{-1}\widehat{(\text{tr}(\nabla u)^2)})(k,s)ds\right] e_k(x),$$

where $e_k = e^{2\pi i k\cdot x}$ are the Fourier components and the A_t^k are the oscillatory Ornstein–Uhlenbeck-type processes (1.35) and $\text{tr}(\nabla u)^2$ denotes the trace of the matrix $(\nabla u)^2$. The Fourier transform of the term $\nabla\Delta^{-1}(\text{tr}(\nabla u)^2)$ is just $\frac{-ik}{2\pi|k|^2}\widehat{\text{tr}(\nabla u)^2}$ and we will write the integral equation in the form

B. Birnir, *The Kolmogorov-Obukhov Theory of Turbulence: A Mathematical Theory of Turbulence*, SpringerBriefs in Mathematics, DOI 10.1007/978-1-4614-6262-0,

$$u(x,t) = \sum_{k\neq 0}\Big[h_k^{1/2}A_t^k - \int_0^t e^{-[4\pi^2\nu|k|^2+2\pi iB(k)](t-s)-2\pi i g(k,t,s)}$$
$$\times\left(\widehat{u\cdot\nabla u}+\frac{ik}{2\pi|k|^2}(\widehat{\mathrm{tr}(\nabla u)^2})\right)(k,s)ds\Big]\, e_k(x), \tag{A.2}$$

where $B(k) = Uk_1 + A(k_2,k_3)\Omega$, from here on with

$$g(k,t,s) = A(k_2,k_3)[\Omega(t-s) - (\sin(\Omega t+\theta) - \sin(\Omega s+\theta))]. \tag{A.3}$$

We will also assume the trivial non-resonance conditions that A and Ω are sufficiently incommensurate for the rest of the chapter.

We split the t integral into the integral from 0 to $t-\delta$, where δ is a small number, and the integral from $t-\delta$ to t. This is done to first avoid the singularities of the spatial derivatives of the heat kernel at $s=t$ and then to deal with these singularities in the latter integral. Now the first estimate is relatively straightforward. The L^2 norm of

$$\sum_{k\neq 0}\int_{t-\delta}^t e^{-\{(4\pi^2\nu|k|^2+2\pi iB)(t-s)+2\pi i g(k,t,s)\}}(-\widehat{u\cdot\nabla u})ds\, e_k$$

is

$$\sum_{k\neq 0}|\int_{t-\delta}^t e^{-\{(4\pi^2\nu|k|^2+2\pi iB)(t-s)+2\pi i g(k,t,s)\}}(-\widehat{u\cdot\nabla u})ds|^2$$
$$\leq \delta\sum_{k\neq 0}\int_{t-\delta}^t |\widehat{u\cdot\nabla u}|_2^2(k)ds$$
$$\leq \delta\int_{t-\delta}^t |u\cdot\nabla u|_2^2 ds \leq \delta \text{ ess sup}_{[t-\delta,t]}|u|_\infty^2\int_{t-\delta}^t |\nabla u|_2^2 ds$$
$$\leq \left(\frac{\delta}{\nu}\int_{t-\delta}^t \langle u, h_k^{1/2}e_k\rangle db_s^k + \frac{\delta^2}{2\nu}\sum_{k\neq 0}h_k\right)\text{ ess sup}_{[t-\delta,t]}\|u\|_{\frac{3}{2}^+}(s) \tag{A.4}$$

since by the Gagliardo–Nirenberg inequalities

$$|u|_\infty \leq C\|u\|_{\frac{3}{2}^+},$$

where δ is independent of U_1 and C is a constant, and by the a priori estimate in Lemma 4.1. Similarly, the L^2 norm of

$$\sum_{k\neq 0}\int_{t-\delta}^t e^{-\{(4\pi^2\nu|k|^2+2\pi iB)(t-s)+2\pi i g(k,t,s)\}}\left(\frac{ik}{2\pi|k|^2}(\widehat{\mathrm{tr}(\nabla u)^2})\right)ds\, e_k$$

is

$$\sum_{k\neq 0}\left|\int_{t-\delta}^{t} \mathrm{e}^{-\{(4\pi^2\nu|k|^2+2\pi \mathrm{i}B)(t-s)+2\pi \mathrm{i}g(k,t,s)\}}\left(\frac{\mathrm{i}k}{2\pi|k|^2}(\widehat{\mathrm{tr}(\nabla u)^2})\right)\mathrm{d}s\right|^2 \tag{A.5}$$

$$\le \delta\sum_{k\neq 0}\int_{t-\delta}^{t}|\left(\frac{\mathrm{i}k}{2\pi|k|^2}(\widehat{\mathrm{tr}(\nabla u)^2})\right)|_2^2(k)\mathrm{d}s$$

$$\le \frac{\delta}{2\pi}\int_{t-\delta}^{t}|\mathrm{tr}(w\cdot\nabla u)|_2^2\mathrm{d}s \;\le\; \frac{\delta}{2\pi}\ \mathrm{ess\ sup}_{[t-\delta,t]}|w|_\infty^2\int_{t-\delta}^{t}|\nabla u|_2^2\mathrm{d}s$$

$$\le \left(\frac{\delta}{2\pi\nu}\int_{t-\delta}^{t}\langle u,h_k^{1/2}e_k\rangle \mathrm{d}b_s^k+\frac{\delta^2}{4\pi\nu}\sum_{k\neq 0}h_k\right)\ \mathrm{ess\ sup}_{[t-\delta,t]}\|u\|_{\frac{3}{2}^+}^2(s), \tag{A.6}$$

where $w=\sum_{k\neq 0}\frac{k}{|k|^2}|k\otimes\hat{u}(k,s)|e_k$, $|w|_2=|u|_2$.

The other integrals are estimated by use of Lemma 4.3. The integral

$$\int_0^{t-\delta}\mathrm{e}^{-\{(4\pi^2\nu|k|^2+2\pi\mathrm{i}B)(t-s)+2\pi\mathrm{i}g(k,t,s)\}}\widehat{u\cdot\nabla u}\mathrm{d}s$$

can be estimated by Lemma 4.3; when $t-\delta$ is an even integer multiple of $\frac{1}{B}$, we get that

$$\int_0^{t-\delta}\mathrm{e}^{-\{4\pi^2\nu|k|^2+2\pi\mathrm{i}B)(t-s)+2\pi\mathrm{i}g(k,t,s)\}}\widehat{u\cdot\nabla u}(s)\mathrm{d}s$$

$$=\frac{1}{2}\int_0^{t-\delta}\Bigg[\mathrm{e}^{-\{4\pi^2\nu|k|^2(t-s)+2\pi\mathrm{i}g(k,t,s)\}}\widehat{u\cdot\nabla u}(s)$$

$$-\mathrm{e}^{-\{4\pi^2\nu|k|^2+2\pi\mathrm{i}B)(t-(s+\frac{1}{2B}))+2\pi\mathrm{i}g(k,t,s+\frac{1}{2B})\}}\widehat{u\cdot\nabla u}\left(s+\frac{1}{2B}\right)\Bigg]\mathrm{e}^{-2\pi\mathrm{i}B(t-s)}\mathrm{d}s$$

$$=\frac{1}{2}\int_0^{t-\delta}\Bigg[\Bigg(\mathrm{e}^{-\{4\pi^2\nu|k|^2(t-s)+2\pi\mathrm{i}g(k,t,s)\}}$$

$$-\mathrm{e}^{-\{4\pi^2\nu|k|^2(t-(s+\frac{1}{2B}))+2\pi\mathrm{i}g(k,t,s+\frac{1}{2B})\}}\Bigg)\widehat{u\cdot\nabla u}(s)\Bigg]\mathrm{e}^{-2\pi\mathrm{i}B(t-s)}\mathrm{d}s$$

$$+\frac{1}{2}\int_0^{t-\delta}\Bigg\{\mathrm{e}^{-(4\pi^2\nu|k|^2\left(t-\left(s+\frac{1}{2B}\right)\right)+2\pi\mathrm{i}g(k,t,s+\frac{1}{2B})\}}\left(\left[\hat{u}(s)-\widehat{u\left(s+\frac{1}{2B}\right)}\right]*\widehat{\nabla u(s)}\right.$$

$$\left.+u\left(s+\frac{1}{2B}\right)*\left[\widehat{\nabla u(s)}-\widehat{\nabla u\left(s+\frac{1}{2B}\right)}\right]\right)\Bigg\}\mathrm{e}^{-2\pi\mathrm{i}B(t-s)}\mathrm{d}s.$$

The first term in the last line above is estimated by Schwarz's inequality

$$\Bigg|\int_0^{t-\delta}\Bigg[\Bigg(\mathrm{e}^{-\{4\pi^2\nu|k|^2(t-s)+2\pi\mathrm{i}g(k,t,s)\}}$$

$$-\mathrm{e}^{-\{4\pi^2\nu|k|^2(t-(s+\frac{1}{2B}))+2\pi\mathrm{i}g(k,t,s+\frac{1}{2B})\}}\Bigg)\widehat{u\cdot\nabla u}(s)\Bigg]\mathrm{e}^{-2\pi\mathrm{i}B(t-s)}\mathrm{d}s\Bigg|^2$$

$$\le\int_0^{t-\delta}\left|\mathrm{e}^{-\{2\pi^2\nu|k|^2(t-s)+2\pi\mathrm{i}g(k,t,s)\}}\right.$$

$$-\mathrm{e}^{-\{2\pi^2\nu|k|^2(t-(s+\frac{1}{B}))+2\pi i g(k,t,s+\frac{1}{2B})\}}\Big|^2 |u|_2^2(s)\,\mathrm{d}s \int_0^{t-\delta} \mathrm{e}^{-4\pi^2\nu|k|^2(t-s)}|\nabla u|_2^2(s)\mathrm{d}s$$

$$\leq \mathrm{e}^{-2\pi^2\nu|k|\delta} \int_0^{t-\delta} \Big|\mathrm{e}^{-\{2\pi^2\nu|k|^2(t-s)+2\pi i g(k,t,s)\}} - \mathrm{e}^{-\{2\pi^2\nu|k|^2(t-(s+\frac{1}{B}))+2\pi i g(k,t,s+\frac{1}{2B})\}}\Big|^2 \mathrm{d}s$$

$$\times \int_0^{t-\delta} \mathrm{e}^{-2\pi^2\nu|k|^2(t-s)}|\nabla u|_2^2(s)\mathrm{d}s \text{ ess sup}_{s\in[0,t-\delta]}|u|_2^2(s)$$

$$\leq \frac{C\mathrm{e}^{-2\pi^2\nu|k|\delta}}{B^2} \int_0^{t-\delta} \mathrm{e}^{-2\pi\nu|k|^2(t-s)}|\nabla u|_2^2(s)\mathrm{d}s \text{ ess sup}_{s\in[0,t-\delta]}|u|_2^2(s)$$

by Lemma 4.3. Similarly the second term is estimated by

$$\left|\int_0^{t-\delta} \mathrm{e}^{-(4\pi^2\nu|k|^2(t-(s+\frac{1}{2B}))+2\pi i g(k,t,s+\frac{1}{2B}))}\left(\left[\hat{u}(s) - \widehat{u\left(s+\frac{1}{2B}\right)}\right] * \widehat{\nabla u(s)}\mathrm{e}^{-2\pi i B(t-s)}\right)\mathrm{d}s\right|^2$$

$$\leq \mathrm{e}^{-4\pi^2\nu|k|^2(\delta-\frac{1}{2B})} \int_0^{t-\delta} \left|u(s) - u\left(s+\frac{1}{2B}\right)\right|_2^2 \mathrm{d}s \int_0^{t-\delta} \mathrm{e}^{-4\pi^2\nu|k|^2 s}|\nabla u|_2^2(s)\mathrm{d}s$$

using the Cauchy–Schwarz inequality both on the convolution and the time integral, and the third term is estimated by

$$\left|\int_0^{t-\delta} \mathrm{e}^{-(4\pi^2\nu|k|^2(t-(s+\frac{1}{2B}))+2\pi i g(k,t,s+\frac{1}{2B}))}\left(\widehat{u\left(s+\frac{1}{2B}\right)} * \left[\widehat{\nabla u(s)} - \widehat{\nabla u\left(s+\frac{1}{2B}\right)}\right]\mathrm{e}^{-2\pi i B(t-s)}\right)\mathrm{d}s\right|^2$$

$$\leq \frac{\mathrm{e}^{-8\pi^2\nu|k|(\delta-\frac{1}{2B})}}{8\nu\pi^2|k|^2} \int_0^{t-\delta} \left|\nabla u(s) - \nabla u\left(s+\frac{1}{2B}\right)\right|_2^2 \mathrm{d}s \text{ ess sup}_{s\in[0,t]}|u|_2^2\left(s+\frac{1}{2B}\right).$$

Now the terms

$$H = \int_0^{t-\delta} \left|u(s) - u\left(s+\frac{1}{2B}\right)\right|_2^2 \mathrm{d}s \int_0^{t-\delta} \mathrm{e}^{-4\pi^2\nu|k|^2 s}|\nabla u|_2^2(s)\mathrm{d}s$$

and

$$K = \int_0^{t-\delta} \left|\nabla u(s) - \nabla u\left(s+\frac{1}{2B}\right)\right|_2^2 \mathrm{d}s \text{ ess sup}_{s\in[0,t]}|u|_2^2\left(s+\frac{1}{2B}\right)$$

are estimated by use of Lemmas 4.2 and 4.6. Thus the a priori bounds on the L^2 norms of u and ∇u and their differences in those two lemmas and in Lemmas 4.1 and 4.7 give the inequality

$$\left|\int_0^{t-\delta} \mathrm{e}^{-\{(4\pi^2\nu|k|^2+2\pi i B)(t-s)+2\pi i g(k,t,s)\}}\widehat{u\cdot\nabla u}(s)\mathrm{d}s\right|^2$$

$$\leq C\mathrm{e}^{-4\pi^2\nu|k|(\delta-\frac{1}{2B})} \text{ ess sup}_{s\in[0,t]}\left(\frac{C}{B^2} + H + K + \mathrm{d}(k)\right),$$

where the terms H and K are estimated in Lemma 4.8 and the expectation of $\mathrm{d}(k)$ vanishes.

Now consider the pressure term. By use of Lemma 4.3, we get that

$$\int_0^{t-\delta} \mathrm{e}^{-\{4\pi^2\nu|k|^2+2\pi \mathrm{i}B)(t-s)+2\pi \mathrm{i}g(k,t,s)\}} \frac{\mathrm{i}k}{2\pi|k|^2}\widehat{\mathrm{tr}(\nabla u)^2}\mathrm{d}s$$
$$= \frac{1}{2}\int_0^{t-\delta}\Big\{\mathrm{e}^{-\{4\pi^2\nu|k|^2(t-s)+2\pi \mathrm{i}g(k,t,s)\}}\frac{\mathrm{i}k}{2\pi|k|^2}\widehat{\mathrm{tr}(\nabla u)^2}(s)$$
$$-\mathrm{e}^{-\{4\pi^2\nu|k|^2(t-(s+\frac{1}{2B}))+2\pi \mathrm{i}g(k,t,s+\frac{1}{2B})\}}\frac{\mathrm{i}k}{2\pi|k|^2}\widehat{\mathrm{tr}(\nabla u)^2}\left(s+\frac{1}{2B}\right)\Big\}\mathrm{e}^{-2\pi \mathrm{i}B(t-s)}\mathrm{d}s$$
$$= \frac{1}{2}\int_0^{t-\delta}\Big\{\mathrm{e}^{-\{4\pi^2\nu|k|^2(t-s)+2\pi \mathrm{i}g(k,t,s)\}}$$
$$-\mathrm{e}^{-\{4\pi^2\nu|k|^2(t-(s+\frac{1}{2B}))+2\pi \mathrm{i}g(k,t,s+\frac{1}{2B})\}}\Big\}\frac{\mathrm{i}k}{2\pi|k|^2}\widehat{\mathrm{tr}(\nabla u)^2}(s)\mathrm{d}s$$
$$+\frac{1}{2}\int_0^{t-\delta}\mathrm{e}^{-\{4\pi^2\nu|k|^2(t-(s+\frac{1}{2B}))+2\pi \mathrm{i}g(k,t,s+\frac{1}{2B})\}}$$
$$\times\frac{\mathrm{i}k}{2\pi|k|^2}\mathrm{tr}\left[\left(\widehat{\nabla u(s)}-\widehat{\nabla u\left(s+\frac{1}{2B}\right)}\right)*\left(\widehat{\nabla u(s)}+\widehat{\nabla u\left(s+\frac{1}{2B}\right)}\right)\right]\mathrm{e}^{-2\pi \mathrm{i}B(t-s)}\mathrm{d}s.$$

The first term in the last expression above is estimated as

$$\Big|\int_0^{t-\delta}\Big\{\mathrm{e}^{-\{4\pi^2\nu|k|^2(t-s)+2\pi \mathrm{i}g(k,t,s)\}}$$
$$-\ \mathrm{e}^{-\{4\pi^2\nu|k|^2(t-(s+\frac{1}{2B}))+2\pi \mathrm{i}g(k,t,s+\frac{1}{2B})\}}\Big\}\frac{\mathrm{i}k}{2\pi|k|^2}\widehat{\mathrm{tr}(\nabla u)^2}(s)\mathrm{d}s\Big|^2$$
$$\le \int_0^{t-\delta}\Big|\mathrm{e}^{-\{2\pi^2\nu|k|^2(t-s)+2\pi \mathrm{i}g(k,t,s)\}}$$
$$-\ \mathrm{e}^{-\{2\pi^2\nu|k|^2(t-(s+\frac{1}{B}))+2\pi \mathrm{i}g(k,t,s+\frac{1}{2B})\}}\Big|^2|w|_2^2(s)\mathrm{d}s\int_0^{t-\delta}\mathrm{e}^{-4\pi\nu|k|^2(t-s)}|\nabla u|_2^2(s)\mathrm{d}s$$
$$\le \mathrm{e}^{-2\pi^2\nu|k|\delta}\int_0^{t-\delta}\Big|\mathrm{e}^{-\{2\pi^2\nu|k|^2(t-s)+2\pi \mathrm{i}g(k,t,s)\}}$$
$$-\ \mathrm{e}^{-\{2\pi^2\nu|k|^2(t-(s+\frac{1}{B}))+2\pi \mathrm{i}g(k,t,s+\frac{1}{2B})\}}\Big|^2\mathrm{d}s\int_0^{t-\delta}\mathrm{e}^{-2\pi\nu|k|^2(t-s)}|\nabla u|_2^2(s)\mathrm{d}s$$
$$\times \mathrm{ess\ sup}_{s\in[0,t-\delta]}|w|_2^2(s)$$
$$\le \frac{C\mathrm{e}^{-2\pi^2\nu|k|\delta}}{B^2}\int_0^{t-\delta}\mathrm{e}^{-2\pi\nu|k|^2(t-s)}|\nabla u|_2^2(s)\mathrm{d}s\ \mathrm{ess\ sup}_{s\in[0,t-\delta]},|u|_2^2(s)$$

where w is the same function as above and by Lemma 4.3. The second term is estimated by

$$\Big|\int_0^{t-\delta}\mathrm{e}^{-\{4\pi^2\nu|k|^2(t-(s+\frac{1}{2B}))+2\pi \mathrm{i}g(k,t,s+\frac{1}{2B})\}}$$
$$\times\frac{\mathrm{i}k}{2\pi|k|^2}\mathrm{tr}\left[\left(\widehat{\nabla u(s)}-\widehat{\nabla u\left(s+\frac{1}{2B}\right)}\right)*\left(\widehat{\nabla u(s)}+\widehat{\nabla u\left(s+\frac{1}{2B}\right)}\right)\right]\mathrm{e}^{-2\pi \mathrm{i}B(t-s)}\mathrm{d}s\Big|^2$$

$$\leq e^{-4\pi^2\nu|k|(\delta-\frac{1}{2B})}\int_0^{t-\delta}\left|\nabla u(s)-\nabla u(s+\frac{1}{2B})\right|_2^2 ds\int_0^{t-(\delta-\frac{1}{2B})}e^{-4\pi^2\nu|k|^2 s}|\nabla u|_2^2(s)ds.$$

Thus

$$\left|\int_0^{t-\delta}e^{-\{(4\pi^2\nu|k|^2+2\pi iB)(t-s)+2\pi i g(k,t,s)\}}\frac{ik}{2\pi|k|^2}\widehat{\mathrm{tr}(\nabla u)^2}ds\right|^2$$
$$\leq Ce^{-4\pi^2\nu|k|(\delta-\frac{1}{2B})}\,\mathrm{ess\ sup}_{s\in[0,t]}\left(\frac{C}{|B(k)|^2}+L+d(k)\right),$$

where the expectation of $d(k)$ vanishes and the term

$$L=\int_0^{t-\delta}|\nabla u(s)-\nabla u(s+\frac{1}{2B})|_2^2 ds\int_0^{t-(\delta-\frac{1}{2B})}e^{-4\pi^2\nu|k|^2 s}|\nabla u|_2^2(s)ds$$

is estimated in Lemma 4.8, again by the a priori bounds on the L^2 norms of u and ∇u and their differences in Lemmas 4.1 and 4.7 and Lemmas 4.2 and 4.6.

When $t-\delta$ is not an even integer multiple of $\frac{1}{B(k)}$ we get the additional terms in Corollary 4.1. However these are estimated exactly as the integrals from $t-\delta$ to t and simply add another term multiplied by δ^2 if we choose $\frac{1}{|B|}=\sup_{k\neq 0}\frac{1}{|B(k)|}<\delta$.

Now we assemble the estimates. Up to terms that vanish when the expectation is taken, the L^2 norm of u is bounded by

$$\begin{aligned}|u|_2^2 &\leq 3\sum_{k\neq 0}h_k|A_t^k|^2\\ &+3\sum_{k\neq 0}\left(\left|\int_0^{t-\delta}e^{-(\{4\pi^2\nu|k|^2+2\pi i[k_1U_1+A(k_2,k_3)\Omega]\}(t-s)+2\pi i g(k,t,s))}\right.\right.\\ &\left.\left.\times(\widehat{u\cdot\nabla u}-\nabla\Delta^{-1}\widehat{(\mathrm{tr}(\nabla u)^2)})(k,s)ds\right|^2\right)+\delta^2 C\mathrm{ess\ sup}_{s\in[t-\delta,t]}\|u\|^2_{\frac{11}{6}^+} \qquad (A.7)\\ &\leq 3\sum_{k\neq 0}h_k|A_t^k|^2\\ &+\sum_{k\neq 0}e^{-4\pi^2\nu|k|(\delta-\frac{1}{2B})}\left[\frac{C'}{|B(k)|^2}+H+K+L\right](s)+\delta^2 C\mathrm{ess\ sup}_{s\in[t-\delta,t]}\|u\|^2_{\frac{11}{6}^+}\\ &\leq 3\sum_{k\neq 0}h_k|A_t^k|^2+C\left(\frac{1}{B^2}+\delta^2\right)\mathrm{ess\ sup}_{s\in[t-\delta,t]}\|u\|^2_{\frac{11}{6}^+}+\frac{C'}{B}\end{aligned}$$

by Lemma 4.8.

We now act on the integral equation (A.2) with the operator $\nabla^{(11/6)^+}$ to estimate the derivative $\nabla^{(11/6)^+}u$

$$\nabla^{(11/6)^+}u(x,t)=\sum_{k\neq 0}\left[(2\pi i|k|)^{(11/6)^+}h_k^{1/2}A_t^k\right.$$

$$- \int_0^t (2\pi \mathrm{i}|k|)^{(11/6)^+} \mathrm{e}^{-[4\pi^2 \nu |k|^2 + 2\pi \mathrm{i} B(k)](t-s) - 2\pi \mathrm{i} g(k,t,s)}$$
$$\times \left(\widehat{u \cdot \nabla u} + \frac{\mathrm{i}k}{2\pi |k|^2} (\widehat{\mathrm{tr}(\nabla u)^2}) \right) (k,s) \mathrm{d}s \Bigg] e_k(x), \tag{A.8}$$

where $B(k)$ and $g(k,t,s)$ are as in (A.2). An estimate similar to (A.8) now gives

$$|\nabla^{(11/6)^+} u|_2^2 \leq 3 \sum_{k \neq 0} (2\pi |k|)^{(11/3)^+} h_k |A_t^k|^2$$
$$+3 \sum_{k \neq 0} \Bigg(\Bigg| \int_0^{t-\delta} |k|^{\frac{11}{6}^+} \mathrm{e}^{-(\{4\pi^2 \nu |k|^2 + 2\pi \mathrm{i}[k_1 U_1 + A(k_2,k_3)\Omega]\}(t-s) + 2\pi \mathrm{i} g(k,t,s))}$$
$$\times (\widehat{u \cdot \nabla u} - \nabla \Delta^{-1} \widehat{(\mathrm{tr}(\nabla u)^2)})(k,s) \mathrm{d}s \Bigg|^2 \Bigg)$$
$$+\delta^{\frac{1}{6}^-} C \mathrm{ess\,sup}_{s \in [t-\delta,t]} \|u\|^2_{\frac{11}{6}^+} \tag{A.9}$$
$$\leq 3 \sum_{k \neq 0} (2\pi |k|)^{(11/3)^+} h_k |A_t^k|^2 + \mathrm{ess\,sup}_{s \in [0,t-\delta]} \left[\frac{C'}{B^2} + H + K + L \right] (s)$$
$$+\delta^{\frac{1}{6}^-} C\, \mathrm{ess\,sup}_{s \in [t-\delta,t]} \|u\|^2_{\frac{11}{6}^+} \tag{A.10}$$
$$\leq 3 \sum_{k \neq 0} (2\pi |k|)^{(11/3)^+} h_k |A_t^k|^2 + C \left(\frac{1}{B^2} + \delta^{\frac{1}{6}^-} \right) \mathrm{ess\,sup}_{s \in [t-\delta,t]} \|u\|^2_{\frac{11}{6}^+} + \frac{C'}{B}$$

again by Lemma 4.8.

Combining the estimates (A.8) and (A.10) we now get that

$$\|u\|^2_{\frac{11}{6}^+} \leq 3 \sum_{k \neq 0} (1 + (2\pi |k|)^{\frac{11}{3}^+}) h_k |A_t^k|^2 + C \left(\frac{1}{B^2} + \delta^{\frac{1}{6}^-} \right) \mathrm{ess\,sup}_{s \in [t-\delta,t]} \|u\|^2_{\frac{11}{6}^+} + \frac{C'}{B},$$

where $\frac{1}{B}$ and δ can be made arbitrarily small. Then taking the expectation we get

$$\left(1 - C \left(\frac{1}{B^2} + \delta^{\frac{1}{6}^-} \right) \right) E(\mathrm{ess\,sup}_{[0,t]} \|u\|^2_{\frac{11}{6}^+}) \leq 3 \sum_{k \neq 0} (1 + (2\pi |k|)^{\frac{11}{3}^+}) h_k E(|A_t^k|^2) + \frac{C'}{B} \tag{A.11}$$

and evaluating the last expectation

$$\sum_{k \neq 0} (1 + (2\pi |k|)^{\frac{11}{3}^+}) h_k E(|A_t^k|^2) = \sum_{k \neq 0} \frac{(1 + (2\pi |k|)^{\frac{11}{3}^+})}{8\pi^2 \nu |k|^2} h_k$$

gives the estimate (4.25)

$$\left(1 - C\left(\frac{1}{B^2} + \delta^{\frac{1}{6}^-}\right)\right) E(\text{ess sup}_{[0,t]} \|u\|^2_{\frac{11}{6}^+}) \leq 3 \sum_{k \neq 0} \frac{(1 + (2\pi|k|)^{\frac{11}{3}^+})}{8\pi^2 \nu |k|^2} h_k + \frac{C'}{B}$$

By making δ and $\frac{1}{B}$ sufficiently small we conclude that (4.25) holds for all t. This ends the outline of the proof; see [17] for more details.

Appendix B
Detailed Estimates of S_2 and S_3

In this appendix we will spell out all the details of the estimates of the structure functions of turbulence S_2 and S_3 from Sect. 2.8. The point is that most researchers may not need the full power of the invariant measures (3.5) and (3.12). Instead they will need the ability to compute averages such as the one that gives us the structure functions in Sect. 2.8, using the expectation $E(\cdot)$ that comes with the noise in the stochastic Navier–Stokes equation (1.66). We will give all the details of this computation below.

We start with the second structure function S_2,

$$E((u(x,t)-u(y,t))^2).$$

The expectation E is actually composed of two expectations; one for the infinite-dimensional Brownian motion and the other for the log-Poisson process.

1. The expectation E_b is that of the Brownian motion b_t^k. This is the integral over $\mathbb{R}$ with the probability density (1.51) and in practice uses the identity (1.63). One evaluates this mean with respect to all the ks, $k = 1, \ldots, \infty$.
2. The second expectation comes from the jumps in the velocity gradients and the multiplicative noise. It is the expectation of the (log) Poisson process N^k

$$E_p(f(N^k)) = \sum_{n=0} f(n)\lambda^n \frac{\mathrm{e}^{-\lambda}}{n!},$$

with the probability (1.45) and the rate λ from (2.26). In practice the function f will be a power.

With these two expectations one can compute the expectation E above, namely,

$$E(\cdot) = E_b \circ E_p(\cdot),$$

where $\circ$ denotes composition of the distributions.

B. Birnir, *The Kolmogorov-Obukhov Theory of Turbulence: A Mathematical Theory of Turbulence*, SpringerBriefs in Mathematics, DOI 10.1007/978-1-4614-6262-0,

We first write the difference

$$u(x,t)-u(y,t)$$
$$=\sum_{k\neq 0}\left[c_k^{1/2}\int_0^t e^{K(t-s)}e^{\int_s^t dq}M_{t-s}db_s^k+d_k\int_0^t e^{K(t-s)}e^{\int_s^t dq}M_{t-s}|k|^{1/3}ds(e_k(x)-e_k(y))\right].$$

The computation of S_2 is now straightforward:

$$\begin{aligned}S_2&=E((u(x,t)-u(y,t))^2)\\&=E\left(\sum_{k\neq 0}\left[c_k^{1/2}\int_0^t e^{K(t-s)}e^{\int_s^t dq}M_{t-s}db_s^k+d_k\int_0^t e^{K(t-s)}e^{\int_s^t dq}M_{t-s}|k|^{1/3}ds(e_k(x)-e_k(y))\right]\right.\\&\quad\left.\times\sum_{k\neq 0}\left[c_k^{1/2}\int_0^t e^{K(t-s)}e^{\int_s^t dq}M_{t-s}db_s^k+d_k\int_0^t e^{K(t-s)}e^{\int_s^t dq}M_{t-s}|k|^{1/3}ds(e_k(x)-e_k(y))\right]\right)\\&=E_b\left(\sum_{k\neq 0}\left[c_k^{1/2}\int_0^t e^{K(t-s)}E_p((|k|^{2/3}(2/3)^{N_t^k})^{2/3})M_{t-s}db_s^k\right.\right.\\&\quad\left.\left.+d_k\int_0^t e^{K(t-s)}E_p((|k|^{2/3}(2/3)^{N_t^k})^{2/3})M_{t-s}|k|^{1/3}ds(e_k(x)-e_k(y))^2\right]\right),\end{aligned}$$

where the application of the expectation E_p selects the "diagonal" $j=k$ from all the products of the Fourier components $(e_k(x)-e_k(y))(e_j(x)-e_j(y))$ and then evaluates the mean $E_p((|k|^{2/3}(2/3)^{N_t^k})^{2/3})$ of the log-Poisson process; see Example 1.5. Now multiplying out the terms in each Fourier component and taking the Brownian motion expectation gives

$$S_2=-\frac{4}{C^2}\sum_{k\in\mathbb{Z}^3}\frac{[d_k^2(1-e^{-\lambda_k t})^2+(C/2)c_k(1-e^{-2\lambda_k t})]}{|k|^{\zeta_2}}e^{2\pi i k\cdot(x+y)}\sin^2(\pi k\cdot(x-y)),$$

since $E_p([|k|^{2/3}2/3^{N_t^k}]^{2/3})=|k|^{-\tau_2}$ by (2.28). We have set $\lambda_k=C|k|^{2/3}$, its "maximum value" without viscosity, and then taking the absolute value we get the estimate

$$S_2\leq\frac{4}{C^2}\sum_{k\in\mathbb{Z}^3}\frac{[d_k^2(1-e^{-\lambda_k t})^2+(C/2)c_k(1-e^{-2\lambda_k t})]}{|k|^{\zeta_2}}\sin^2(\pi k\cdot(x-y)).$$

The computation of S_3 is similar:

$$\begin{aligned}S_3&=E((u(x,t)-u(y,t))^2)\\&=E\left(\left(\sum_{k\neq 0}\left[c_k^{1/2}\int_0^t e^{K(t-s)}e^{\int_s^t dq}M_{t-s}db_s^k+d_k\int_0^t e^{K(t-s)}e^{\int_s^t dq}M_{t-s}|k|^{1/3}ds(e_k(x)-e_k(y))\right]\right)^3\right)\\&=E_b\left(\sum_{k\neq 0}\left[c_k^{1/2}\int_0^t e^{K(t-s)}E_p(|k|^{2/3}(2/3)^{N_t^k})M_{t-s}db_s^k\right.\right.\\&\quad\left.\left.+d_k\int_0^t e^{K(t-s)}E_p(|k|^{2/3}(2/3)^{N_t^k})M_{t-s}|k|^{1/3}ds(e_k(x)-e_k(y))\right]^3\right)\end{aligned}$$

$$= \frac{2^3}{C^3} \sum_{k \neq 0} \frac{[\mathrm{d}_k^3 (1-\mathrm{e}^{-\lambda_k t})^2 + (C/2) c_k \mathrm{d}_k (1-\mathrm{e}^{-2\lambda_k t})(1-\mathrm{e}^{-\lambda_k t})]}{|k|} \mathrm{e}^{3\pi i k\cdot(x+y)} \sin^3(\pi k \cdot (x-y)),$$

since $E_p(|k|^{2/3} 2/3^{N_t^k}) = 1$; see Example 1.5. Thus we get the estimate

$$S_3 \leq \frac{2^3}{C^3} \sum_{k \neq 0} \frac{[|\mathrm{d}_k|^3 (1-\mathrm{e}^{-\lambda_k t})^2 + (C/2) c_k |\mathrm{d}_k| (1-\mathrm{e}^{-2\lambda_k t})(1-\mathrm{e}^{-\lambda_k t})]}{|k|} |\sin^3(\pi k \cdot (x-y))|.$$

The estimate of the higher-order structure functions is similar.

Appendix C
The Generalized Hyperbolic Distributions

The generalized hyperbolic distribution (GHD) was defined by Barndorff-Nielsen [6] to explain the size distribution of windblown grains of sand. Its probability density function is given by the formula

$$H(\lambda,\alpha/a,\beta/a,a\delta,a\mu+b) = \frac{\gamma^{\lambda}}{\sqrt{2\pi}\delta^{\lambda}K_{\lambda}(\delta\gamma)}\frac{K_{\lambda-1/2}(\alpha\sqrt{\delta^2+(x-\mu)^2})}{(\sqrt{\delta^2+(x-\mu)^2}/\alpha)^{\lambda-1/2}}e^{\beta(x-\mu)}, \tag{C.1}$$

where $\gamma = \sqrt{\alpha^2-\beta^2}$, $K_{\lambda-1/2}$ and K_{λ} are modified Bessel's function of the second kind with index $\lambda - 1/2$ and λ, respectively. In (3.19) we set $\lambda = 3/2$. This is a five-parameter distribution, if one counts the index λ with the four parameters $(\alpha,\beta,\delta,\mu)$; recall that $\gamma = \sqrt{\alpha^2-\beta^2}$. The ranges for the parameters are $\lambda \in \mathbb{R}$, $\alpha > 0$, $\beta \in (-\alpha,\alpha)$, $\delta > 0$ and $\mu \in \mathbb{R}$.

The moment-generating function for the GHD is

$$M_{(\lambda,\alpha,\beta,\delta,\mu)}(z) = e^{\mu z}\frac{(\delta\gamma)^{\lambda}}{K_{\lambda}(\delta\gamma)}\frac{K_{\lambda}(\delta\sqrt{\alpha^2-(\beta+z)^2})}{(\delta\sqrt{\alpha^2-(\beta+z)^2})^{\lambda}}. \tag{C.2}$$

Using M we immediately compute the mean and the variance

$$E(X) = \mu + \frac{\delta\beta K_{\lambda+1}(\delta\gamma)}{\gamma K_{\lambda}(\delta\gamma)} \tag{C.3}$$

and

$$\mathrm{var}(X) = \frac{\delta K_{\lambda+1}(\delta\gamma)}{\gamma K_{\lambda}(\delta\gamma)} + \frac{\beta^2\delta^2}{\gamma^2}\left(\frac{K_{\lambda+2}(\delta\gamma)}{K_{\lambda}(\delta\gamma)} - \frac{K^2_{\lambda+1}(\delta\gamma)}{K^2_{\lambda}(\delta\gamma)}\right) \tag{C.4}$$

of the generalized hyperbolic random variable X.

The class of GHDs is closed under affine transformations. That is, if $X \sim H(\lambda,\alpha,\beta,\delta,\mu)$ and $Y = aX+b$ for some positive a, we have

$$Y \sim H(\lambda,\alpha/a,\beta/a,a\delta,a\mu+b),$$

B. Birnir, *The Kolmogorov-Obukhov Theory of Turbulence: A Mathematical Theory of Turbulence*, SpringerBriefs in Mathematics, DOI 10.1007/978-1-4614-6262-0,

where $X \sim H$ means that Y has the distribution H. It is sometime useful to parametrize the GHD in terms of the parameters λ, τ, ζ, δ, and μ, where $\tau = \beta/\gamma$, and $\zeta = \delta\gamma$. The first three parameters λ, τ and ζ are invariant under an affine transformation of X. In other words

$$Y \sim H(\lambda, \tau, \zeta, a\delta, a\mu + b),$$

where $Y = aX + b$ as above. This shows that δ is a scaling parameter and μ centers (or locates) the distribution. The parameters $\chi = \frac{\beta/\alpha}{\sqrt{1+\zeta}}$ and $\xi = \frac{1}{\sqrt{1+\zeta}}$ are known as the skewness and kurtosis parameters for the shape triangle of the GHD. Since $0 \leq |\chi| \leq \xi < 1$, the GHD parametrized by χ and ξ, can be represented by points on a triangle, called the shape triangle; see [8].

The characteristic function is obtained from (C.2) using the relation

$$\phi_X(h) = M(\mathrm{i}h).$$

The Lévy–Khintchine representation of the characteristic function of GHDs is given by

$$\ln(\phi_X(h)) = \mathrm{i}hE(X) + \int_{-\infty}^{\infty} (\mathrm{e}^{\mathrm{i}hx} - 1 - \mathrm{i}hx)g(x)\mathrm{d}x, \tag{C.5}$$

where $g(x)$ is the density of the Lévy measure

$$g(x) = \frac{\mathrm{e}^{\beta x}}{|x|}\left(\int_0^{\infty} \frac{\mathrm{e}^{-\sqrt{2y+\alpha^2}|x|}}{\pi^2 y(J_\lambda^2(\delta\sqrt{2y}) + Y_\lambda^2(\delta\sqrt{2y}))}\mathrm{d}y + \lambda \mathrm{e}^{-\alpha|x|}\right), \qquad \text{if } \lambda \geq 0,$$

and

$$g(x) = \frac{\mathrm{e}^{\beta x}}{|x|}\left(\int_0^{\infty} \frac{\mathrm{e}^{-\sqrt{2y+\alpha^2}|x|}}{\pi^2 y(J_{-\lambda}^2(\delta\sqrt{2y}) + Y_{-\lambda}^2(\delta\sqrt{2y}))}\mathrm{d}y\right), \qquad \text{if } \lambda < 0,$$

where J_λ and Y_λ are Bessel's function of the first and the second kind, respectively. The NIG distribution is a GHD with the parameter value $\lambda = 1$.

References

1. R. A. Adams. *Sobolev Spaces*. Academic Press, New York, 1975.
2. F. Anselmet, Y. Gagne, E. J. Hopfinger, and R. A. Antonia. High-order velocity structure function sin turbulent shear flows. *J. Fluid Mech.*, 14:63–89, 1984.
3. A. Babin, A. Mahalov, and B. Nicolaenko. Long-time averaged Euler and Navier-Stokes equations for rotation fluids. *In Structure and Dynamics of non-linear waves in Fluids, 1994 IUTAM Conference, K. Kirehgassner and A. Mielke (eds), World Scientific*, page 145–157, 1995.
4. A. Babin, A. Mahalov, and B. Nicolaenko. Global splitting, integrability and regularity of 3d Euler and Navier-Stokes equation for uniformly rotation fluids. *Eur. J. Mech. B/Fluids*, 15(2):08312, 1996.
5. A. V. Babin and M. I Vishik. *Attractors of Evolution Equations*. Studies in Appl. Math and its Applic. vol. 25, North Holland Amsterdam, 1992.
6. O. E. Barndorff-Nielsen. Exponentially decreasing distributions for the logarithm of the particle size. *Proc. R. Soc. London*, A 353:401–419, 1977.
7. O. E. Barndorff-Nielsen. Processes of normal inverse Gaussian type. *Finance and Stochastics*, 2:41–68, 1998.
8. O. E. Barndorff-Nielsen, P. Blaesild, and Jurgen Schmiegel. A parsimonious and universal description of turbulent velocity increments. *Eur. Phys. J. B*, 41:345–363, 2004.
9. G. K. Batchelor. *The Theory of Homogenous Turbulence*. Cambridge Univ. Press, New York, 1953.
10. P. S. Bernard and J. M. Wallace. *Turbulent Flow*. John Wiley & Sons, Hoboken, NJ, 2002.
11. R. Betchov and W. O. Criminale. *Stability of Parallel Flows*. Academic Press, New York, 1967.
12. R. Bhattacharya and E. C. Waymire. *Stochastic Processes with Application*. John Wiley, New York, 1990.
13. R. Bhattacharya and E. C. Waymire. *A Basic Course in Probability Theory*. Springer, New York, 2007.
14. P. Billingsley. *Probability and Measure*. John Wiley, New York, 1995.
15. B. Birnir. Turbulence of a unidirectional flow. *Proceedings of the Conference on Probability, Geometry and Integrable Systems, MSRI, Dec. 2005 MSRI Publications, Cambridge Univ. Press*, 55, 2007. Available at `http://repositories.cdlib.org/cnls/`.
16. B. Birnir. Turbulent Rivers. *Quarterly of Applied Mathematics*, 66:565–594, 2008.
17. B. Birnir. The Existence and Uniqueness and Statistical Theory of Turbulent Solution of the Stochastic Navier-Stokes Equation in three dimensions, an overview. *Banach J. Math. Anal.*, 4(1):53–86, 2010. Available at `http://repositories.cdlib.org/cnls/`.
18. B. Birnir. The Kolmogorov-Obukhov statistical theory of turbulence. To appear in *Journal of Nonlinear Science*, 2013. Available at `http://repositories.cdlib.org/cnls/`.

B. Birnir, *The Kolmogorov-Obukhov Theory of Turbulence: A Mathematical Theory of Turbulence*, SpringerBriefs in Mathematics, DOI 10.1007/978-1-4614-6262-0,

19. S. Y. Chen, B. Dhruva, S. Kurien, K. R. Sreenivasan, and M. A. Taylor. Anomalous scaling of low-order structure functions of turbulent velocity. *Journ. of Fluid Mech.*, 533:183–192, 2005.
20. P. A. Davidson, Y. Kaneda, K. Moffatt, and K. R. Sreenivasan. *A Voyage Through Turbulence*. Cambridge Univ. Press, New York, 2012.
21. A. Debussche and C. Odasso. Markov solutions for the 3 d stochastic Navier-Stokes equations with state dependent noise. *Journal of Evolution Equations*, 6(2):305–324, 2006.
22. B. Dhruva. *An experimental study of high-Reynolds-number turbulence in the atmosphere*. Ph.D. Thesis Yale University, New Haven, CT, 2000.
23. B. Dubrulle. Intermittency in fully developed turbulence: in log-Poisson statistics and generalized scale covariance. *Phys. Rev. Letters*, 73(7):959–962, 1994.
24. F. Flandoli and G. Gatarek. Martingale and stationary solutions for stochastic Navier-Stokes equations. *Prob. Theory Rel. Fields*, 102:367–391, 1995.
25. C. Foias, O. Manley, R. Rosa, and R. Temam. *Navier-Stokes Equations and Turbulence*. Cambridge Univ. Press, Cambridge UK, 2001.
26. U. Frisch. *Turbulence*. Cambridge Univ. Press, Cambridge, 1995.
27. M. Hairer and J. Mattingly. Ergodic properties of highly degenerate 2d stochastic Navier-Stokes equations. *Comptes Rendus Mathématique. Académie des Sciences. Paris*, 339(12):879–882, 2004.
28. M. Hairer, J. Mattingly, and É. Pardoux. Malliavin calculus for highly degenerate 2d stochastic Navier-Stokes equations. *Comptes Rendus Mathématique. Académie des Sciences. Paris*, 339(11):793–796, 2004.
29. E. Hopf. Statistical hydrodynamics and functional calculus. *J. Rat. Mech. Anal.*, 1(1):87–123, 1953.
30. S. Hou, B. Birnir, and N. Wellander. Derivation of the viscous Moore-Greitzer equation for aeroengine flow. *Journ. Math. Phys.*, 48:065209, 2007.
31. B.R. Hunt, T. Sauer, and J.A. Yorke. Prevalence: A translation-invariant "almost every" on infinite-dimensional spaces. *Bull. of the Am. Math. Soc.*, 27(2):217–238, 1992.
32. T. Kato. *Perturbation Theory for Linear Operators*. Springer, New York, 1976.
33. J. F. C. Kingman. *Poisson Processes*. Clarendon Press, Oxford, 1993.
34. A. N. Kolmogorov. Dissipation of energy under locally isotropic turbulence. *Dokl. Akad. Nauk SSSR*, 32:16–18, 1941.
35. A. N. Kolmogorov. The local structure of turbulence in incompressible viscous fluid for very large Reynolds number. *Dokl. Akad. Nauk SSSR*, 30:9–13, 1941.
36. A. N. Kolmogorov. A refinement of previous hypotheses concerning the local structure of turbulence in a viscous incompressible fluid at high Reynolds number. *J. Fluid Mech.*, 13:82–85, 1962.
37. R. H. Kraichnan. The structure of isotropic turbulence at very high Reynolds numbers. *J. Fluid Mech.*, 5:497–543, 1959.
38. R. H. Kraichnan. Lagrangian-history closure approximation for turbulence. *Phys. Fluids*, 8:575–598, 1965.
39. R. H. Kraichnan. Inertial ranges in two dimensional turbulence. *Phys. Fluids*, 10:1417–1423, 1967.
40. R. H. Kraichnan. Turbulent cascade and intermittency growth. *In Turbulence and Stochastic Processes, eds. J. C. R. Hunt, O. M. Phillips and D. Williams, Royal Society*, pages 65–78, 1991.
41. S. Kuksin and A. Shirikyan. A coupling approach to randomly forced nonlinear pdes. *Comm. Math. Phys.*, 221:351–366, 2001.
42. J. Leray. Sur le mouvement d'un liquide visqueux emplissant l'espace. *Acta Math.*, 63(3):193–248, 1934.
43. J. Lunch and J. Sethuraman. Large deviations for processes with independent increments. *The Annals of Probability*, 15(2):610–627, 1987.
44. H. P. McKean. Turbulence without pressure: Existence of the invariant measure. *Methods and Applications of Analysis*, 9(3):463–468, 2002.

45. J. Milnor. On the concept of attractor. *Communications in Mathematical Physics*, 99:177–195, 1985.
46. A. S. Momin and A. M. Yaglom. *Statistical Fluid Mechanics*, volume 1. MIT Press, Cambridge, MA, 1971.
47. A. S. Momin and A. M. Yaglom. *Statistical Fluid Mechanics*, volume 2. MIT Press, Cambridge, MA, 1975.
48. M. Nelkin. Turbulence in fluids. *Am. J. Phys.*, 68(4):310–318, 2000.
49. A. M. Obukhov. On the distribution of energy in the spectrum of turbulent flow. *Dokl. Akad. Nauk SSSR*, 32:19, 1941.
50. A. M. Obukhov. Some specific features of atmospheric turbulence. *J. Fluid Mech.*, 13:77–81, 1962.
51. B. Oksendal. *Stochastic Differential Equations*. Springer, New York, 1998.
52. B. Oksendal and A. Sulem. *Applied Stochastic Control of Jump Diffusions*. Springer, New York, 2005.
53. L. Onsager. The distribution of energy in turbulence. *Phys. Rev.*, 68:285, 1945.
54. L. Onsager. Statistical hydrodynamics. *Nuovo Cimento.*, 6(2):279–287, 1945.
55. S. B. Pope. *Turbulent Flows*. Cambridge Univ. Press, Cambridge UK, 2000.
56. G. Da Prato. *An Introduction of Infinite-Dimensional Analysis*. Springer Verlag, New York, 2006.
57. G. Da Prato and J. Zabczyk. *Stochastic Equations in Infinite Dimensions*. Cambridge University Press, Cambridge UK, 1992.
58. G. Da Prato and J. Zabczyk. *Ergodicity for Infinite Dimensional Systems*. Cambridge University Press, Cambridge UK, 1996.
59. R. Renzi, S. Ciliberto, C. Baudet, F. Massaioli, R. Tripiccione, and S. Succi. Extended self-similarity in turbulent flow. *Phys. Rev. E*, 48(29):401–417, 1993.
60. O. Reynolds. An experimental investigation of the circumstances which determine whether the motion of water shall be direct or sinuous, and the law resistance in parallel channels. *Phil. Trans. Roy. soc. Lond.*, 174(11):935–982, 1883.
61. R.Temam. *"Infinite-Dimensional Dynamical Systems in Mechanics and Physics"*. Springer New York, 1988.
62. D. Ruelle. "Large volume limit of distribution of characteristic exponents in turbulence". *Comm. Math. Phys.*, 87:287–302, 1982.
63. D. Ruelle. "Characteristic exponents for a viscous fluid subjected to time-dependent forces". *Comm. Math. Phys.*, 92:285–300, 1984.
64. Z-S She and E. Leveque. Universal scaling laws in fully developed turbulence. *Phys. Rev. Letters*, 72(3):336–339, 1994.
65. Z-S She and E. Waymire. Quantized energy cascade and log-poisson statistics in fully developed turbulence. *Phys. Rev. Letters*, 74(2):262–265, 1995.
66. Z-S She and Zhi-Xiong Zhang. Universal hierarchical symmetry for turbulence and general multi-scale fluctuation systems. *Acta Mech Sin*, 25:279–294, 2009.
67. Y. Sinai. Burgers equation driven by a periodic stochastic flow. *Ito's Stochastic Calculus and Probability Theory, Springer New York*, pages 347–353, 1996.
68. K. R. Sreenivasan and R. A. Antonia. The phenomenology of small-scale turbulence. *Annu. Rev. Fluid Mech.*, 29:435–472, 1997.
69. K. R. Sreenivasan and B. Dhruva. Is there scaling in high- Reynolds-number turbulence? *Prog. Theor. Phys. Suppl.*, 103–120, 1998.
70. G. I. Taylor. Statistical theory of turbulence. *Proc. Royal Soc. London*, 151:421–444, 1935.
71. A. A. Townsend. The passage of turbulence through wire gauzes. *Quart. J. Mech. Appl. Math.*, 4:308–320, 1951.
72. A. A. Townsend. *The Structure of Turbulent Flow*. Cambridge Univ. Press, New York, 1976.
73. S. R. S. Varadhan. *Large Deviations and Applications*. SIAM, Philadelphia, PA, 1884.
74. M. I. Vishik and A. V. Fursikov. *Mathematical Problems of Statistical Hydrodynamics*. Kluwer, Dordrecht, Netherlands, 1988.

75. J. B. Walsh. *An Introduction to Stochastic Differential Equations*. Springer Lecture Notes, eds. A. Dold and B. Eckmann, Springer, New York, 1984.
76. M. Wilczek. *Statistical and Numerical Investigations of Fluid Turbulence*. PhD Thesis, Westfälische Wilhelms Universität, Münster, Germany, 2010.
77. M. Wilczek, A. Daitche, and R. Friedrich. On the velocity distribution in homogeneous isotropic turbulence: correlations and deviations from Gaussianity. *J. Fluid Mech.*, 676:191–217, 2011.
78. H. Xu, N. T. Ouellette, and E. Bodenschatz. Multifractal dimension of Lagrangian turbulence. *Phys. Rev. Letters*, 96:114503, 2006.
79. Y. Yang and D. I. Pullin. Geometric study of Lagrangian and Eulerian structures in turbulent channel flow. *J. Fluid Mech.*, 674:6792, 2011.
80. Y. Yang, D. I. Pullin, and I. Bermejo-Moreno. Multi-scale geometric analysis of Lagrangian structures in isotropic turbulence. *J. Fluid Mech.*, 654:233270, 2010.

Index

B. Birnir, *The Kolmogorov-Obukhov Theory of Turbulence: A Mathematical Theory of Turbulence*, SpringerBriefs in Mathematics, DOI 10.1007/978-1-4614-6262-0,

MIX
Papier aus verantwortungsvollen Quellen
Paper from responsible sources
FSC® C105338

Printed by Books on Demand, Germany